Ben Miller studied physics at Cambridge and was working on his PhD when he left to pursue a career in comedy. He is best known as half of Britain's most popular TV comedy duo, Armstrong & Miller, but has always maintained a passionate interest in science.

IT'S NOT
ROCKET
SCIENCE

Ben Miller

sphere

SPHERE

First published in Great Britain in 2012 by Sphere

A CIP catalogue record for this book
is available from the British Library.

ISBN 978-1-84744-501-8

Typeset in Bembo by M Rules
Printed and bound in Great Britain by
Clays Ltd, St Ives Plc

Papers used by Sphere are from well-managed forests
and other responsible sources.

MIX
Paper from
responsible sources
FSC® C104740

Sphere
An imprint of
Little, Brown Book Group
100 Victoria Embankment
London EC4Y 0DY

An Hachette UK Company
www.hachette.co.uk

www.littlebrown.co.uk

For my father, who gave me my love of science

CONTENTS

CHAPTER 1

FIRST LOVE RUNS DEEP

THE BEGINNING

Did you know that we are all stars? I don't mean that in a Simon Cowell-type, doing-it-for-my-dead-nan way. I mean that people, real people, are quite literally made from stardust. It sounds like the most ridiculous sort of science fiction – but this is the world we live in, seen through the eyes of science.

Let me explain. You, as well as everything around you, are made from atoms. You can think of them as the basic building blocks of nature. There was probably a chart on the wall at school called the Periodic Table that showed them in order of increasing size: the smallest, like hydrogen and helium, up at the top, and the big boys, like lead and uranium, down at the bottom. You'll probably also have a vague recollection that these atoms were themselves made up of even smaller parts; to be precise, there was a small, dense, electrically positive nucleus at the centre, surrounded by a swarm of negatively charged electrons. Well, have you ever wondered how those atoms were made?

The answer, incredible as it may seem, is that they were made inside stars. The reason that the stars shine is that there is an enormous nuclear reaction going on inside them, where smaller nuclei are fused together into bigger nuclei, releasing huge amounts of energy as heat and light. The bigger the star, the bigger the nuclei it can make. And once you've got a nucleus, all you need to do is sprinkle on a few electrons – which, frankly, are ten a penny – and you've got a beautiful, life-giving, electrically neutral, works-straight-out-of-the-box atom.

A star like our Sun, it turns out, is a little on the small side. This means it can produce only the smaller sorts of atoms, like helium. Bigger stars are capable of producing much bigger atoms, like iron and carbon – the sort of stuff that you and I are made of. So how do these bigger atoms get from the inside of a star into our bodies?

The answer is that the life cycle of big stars ends with what Noel Gallagher calls a supernova, a huge explosion that flings debris out across the galaxy. Over billions of years, this debris slowly clumps together due to gravity, sometimes forming new stars, sometimes forming planets. On these planets, if the conditions are right, life can form.

In other words, the atoms that make up our bodies were formed billions of years ago in the centres of real, genuine, 100 per cent stars. Those stars then exploded in ruddy great explosions, the debris became planets, life began on our planet – and then, thanks to a particularly slack period in popular music, Simon Cowell evolved. That's science. It's big, it's bold, and – according to every experimental test we've been able to make – it's true. If that is the sort of thing that gets you going, this is the book for you.

MR BAILEY

I've always loved the arts as well as the sciences, and I've always found it strange that the two subjects are separated by some weird sort of educational apartheid. If you were to generalise about where we stand on this issue today – and what the heck else is a book like this for? – you'd say that the arts have an aristocratic, high-church, Royalist air about them, while the sciences seem altogether more egalitarian, plain-speaking and Puritan. We seem to find ourselves on one side or the other of these cultural tracks, centrally cast either as foppish, airy-fairy creative types or unwashed, unapproachable, socially challenged geeks.

Needless to say, this schism is a very modern invention. For a kick-off, no king was ever more loving of the sciences than that lovingly restored fop-to-end-all-fops, Charles II – and, conversely, it's hard to imagine someone less likely to dissect a frog or launch a weather balloon than Oliver Cromwell. Yet the entire education system seems to buy into the contemporary myth that we are all either artists or scientists from birth.

Can it really be that there are two types of human intelligence, one ideally suited to composing haikus and the other perfect for mucking about with a chemistry set? Why does science become such a passion for a few and yet such a mystery for so many?

I think a lot depends on your earliest experiences of science, and I was extremely fortunate in having one of the finest teachers of natural philosophy that anyone could hope for. His name was Mr Bailey, and the things that he taught me and my fellow infants at Willaston County Primary School have stayed with me throughout my entire adult life. If you'll let me, I'd like to tell you a little bit about how, under his influence, I came to study the sciences.

There is not a lot to say about Willaston, the village where I spent my first few years. There were about half a dozen shops, which all appeared to sell newspapers; a railway level-crossing, which provided the lion's share of the local entertainment; some concrete playing fields with generations of children's broken teeth embedded in the tarmac; and a large housing estate where several hundred young families, of which we were one, clung to the lowest rung of the Cheshire housing ladder.

Willaston County Primary School had been recently built to cater for this sprawl of cookie-cutter housing, and was 'modern' – which, in 1971, basically meant that it had a flat roof. If you haven't grasped it yet, I'm trying to paint a picture of a thoroughly ordinary state school, of the kind that can be found anywhere in the country, unremarkable in every way. Except that, in my opinion, what was going on in the classrooms of Willaston County Primary was anything but ordinary, and that was largely due to our rather unconventional Deputy Headmaster.

Mr Bailey was an unlikely schoolteacher. He cut quite a dash, tall and thin, with pepper-grey hair, a neat moustache and a sartorial mien that hovered just the right side of Basil Brush. He was boundlessly enthusiastic, loved a field trip and practically lived for the opportunity to tell anecdotes. But, of all his many passions, one reigned supreme: maths.

Maths, Mr Bailey said, was just about the most fun you could have next to British Bulldog. Practically the first thing he introduced us to was number bases. The non-mathematicians among you might think you know nothing about number bases – but, of course, you do. In fact, you are absolute experts in one of them: base ten. As Mr Bailey explained, the reason we count to ten, and then in multiples of ten, is because we have ten fingers. But why stop there, asked Mr Bailey? For fun, why not count in base eight, as if we were Mickey Mouse and had only eight fingers? Why not count in base sixteen?

The point being that, from our very first encounter with numbers, we were encouraged to see them as things we could play with. In fact, the home-made woodblock number base sets that Mr Bailey provided for our infant school were just as popular as the Lego or the sandpit. To Mr Bailey, numbers were more than just a necessary evil; they were entertainment. And, though he can't have known it at the time, a knowledge of number bases was going to be very useful to a generation of children whose computers would be designed to run on base two, or, as we now call it, digital.

The rite of passage for any child in one of Mr Bailey's classes was the day you qualified for your tables licence. This was done just as if you were taking a driving test, with two chairs placed side by side. The 'examiner' – one of your classmates – sat next to you and quizzed you on your multiplication tables. If you 'passed' by getting them all correct, a serious little ceremony would take place whereby an official-looking booklet would be signed and countersigned, and your photo glued haphazardly on the cover. The small print declared that 'Ben Miller, the undersigned, has hereby been judged by Philip Buckley to be proficient in times tables one to thirteen, and is now free to use them in perpetuity as he sees fit.' Occasionally, Mr Bailey would make spot checks.

'Excuse me, young man. I see you are multiplying numbers together. Have you got your tables licence?'

The usual fumbling about looking for the vital document would follow. 'Here it is, sir!'

Mr Bailey would study it like an overzealous border guard. 'Very good! Carry on.'

One of Mr Bailey's favourite touchstones was the louche, devil-may-care attitude of the numerate. 'Mathematicians', Mr Bailey would say, 'are lazy.' By his reckoning, the arithmetical profession can't be bothered to add up lots of numbers, so they multiply them instead. After all, who wants to add up eleven fours when you can just use the eleven times table – provided you have the appropriate documentation, of course – and get the answer straight away? And why learn multiplication tables for all numbers when you can get away with learning only the tables for the first thirteen? And had we heard that grown-up mathematicians were so lazy that they had published whole books of answers to sums and called them logarithmic tables?

Willaston County Primary was one of a dozen or so feeder schools for Malbank Secondary School, the comprehensive in nearby Nantwich where I ended up taking my O and A levels. We were a streamed comprehensive, where pupils were grouped according to ability. And this is the bit that makes me wonder how much of what we think is innate about our proficiency in maths or science or anything else depends solely on whether we happen to come across a particularly gifted primary-school teacher. After an especially gruelling set of tests, assessments and best-of-three arm wrestles, I was put in the top stream for maths. There were about thirty-five of us in all. I needn't have worried about making friends; nearly every other child in the class was from Willaston. Needless to say, they had all been taught by Mr Bailey.

THE IDLER

Now I look back on it, one of the main reasons that I chose to study the sciences was laziness. After all, one question about, say, gravitation is pretty much like another; all you need is a few general principles and you're off to the races. Most importantly, no one expects you to remember anything; these days, they even print the formulae on the exam papers to save you the effort of scratching them into your shatterproof ruler.

Not so in the arts, of course. When I finally joined the sixth form, I signed up for A levels in English, History and French, and I got the shock of my life. Everyone talks about learning lessons from history, but we all know that's just a bit of empty salesmanship; sign up for an A level and you are staring down the wrong end of four millennia of random events with dates, times and places to match. There's no shape or form to any of it, frankly, and to make it worse half the kings of England had the same ruddy name. Add to that a Matterhorn of novels from the English reading list and an epidemic of irregular French verbs and you've really got your work cut out.

Needless to say, I beat a hasty retreat to the science block and threw myself upon the mercy of those trusty stalwarts, maths, physics and chemistry. The other thing that lured me back to the fold was, of course, the goodies that awaited at degree level. I was under no illusion about the privations of undergraduate English; my father was a lecturer in English Literature at what was then Birmingham Polytechnic and I had sat in on enough of his classes to know that you didn't spend your time reading out loud from *Fanny Hill*. It seemed to me, in fact, that the further upstream you swam with that particular school, the less it became about character and story, and the more it became about, well, just about anything else: sociology, feminism, Marxism – take your pick.

In science, on the other hand, we knew that our A levels were the grunt work and that the real treats lay beyond – impossibly glamorous things like Relativity and Quantum Theory. If, like me, you had no game plan other than to delay for as long as possible the moment when you might have to actually work for a living, the prospect of a degree in science held very great lustre indeed. Increasingly, for me, a degree in science was coming to mean a degree in physics. And if you wanted to study physics there was only one place to do it: Cambridge.

Why Cambridge? Well, frankly, that's a bit like asking someone why they want to record their album at Abbey Road. Cambridge is to physics what Madison Square Garden is to Simon and Garfunkel. It was at Cambridge that Isaac Newton formulated his theories of optics, motion and gravitation, as well as discovering calculus, a mathematical tool that remains at the heart of practically every equation in modern physics. The famous physics laboratory, the Cavendish, was founded there by one James Clerk Maxwell, arguably the world's most influential physicist after Newton and Einstein, and the first to discover the interconnection of electricity, magnetism and light. And, even more importantly for a spotty teenager in 1985, Cambridge was where Stephen Hawking currently lived.

I had seen a BBC documentary about Hawking and his work on black holes while I was still a sixth former, and it seemed beyond comprehension that I might somehow be permitted to study under the same roof as this god-like figure – or at least, if not under the same roof, in the same city. At that time, Hawking held the Lucasian Chair in Mathematics, the same position that Isaac Newton had occupied, and his reputation was growing by the year. He was on the brink of publishing the best-selling book, *A Brief History of Time*, and was making extraordinary progress in updating General Relativity by combining it with quantum mechanics; that success, together with his battle with motor neurone disease, had made him a strikingly prominent media figure. The documentary I had seen had showed him working closely with a wide range of young graduate students, cajoling and inspiring them as they worked together on theories at the very forefront of human knowledge. Could it be possible that this was what my future held?

Would I end up as one of Stephen Hawking's right-hand men, helping him solve the deepest mysteries of the Universe?

COOL FOR CATZ

The Cambridge interview is not to be taken lightly. I had two. The first was with a very pleasant Admissions Tutor, Dr Carl Baron. He asked me why, with my extraordinarily brilliant A level results, did I not wish to study to become a doctor or a vet? I can only assume he was looking at someone else's application form; with the mixed bag of As, Bs and Cs that I had, I'd have been lucky to become a chiropractor. I mumbled something about how much I loved physics and how the sight of blood would quite possibly make me retch.

My academic interview was, to say the least, hit and miss. I should explain that you can't actually study physics at Cambridge; in fact, you can't study any solo science subject. Everyone applies for one enormous mash-up called Natural Sciences, the idea being that you get a broad grounding in a number of sciences before specialising in your final year. I'd already registered my interest in physics, so my interview was with Dr John Shakeshaft, a physicist, and Dr Paul Raithby, a chemist. It started badly. Dr Shakeshaft asked me a very simple question about Newton's second law; I freaked out under the pressure and I came out with the only thing I could think of: 'We didn't do that at my school.' This is a bit like saying that no one taught you to shampoo at Vidal Sassoon. He tried another tack and asked me something about electromagnetism; I claimed never to have heard of that either. Out of sympathy, I think, Paul Raithby took two objects from his pocket and laid them on the table in front of me: a squash ball and a piece of hard plastic. 'If I told you', he said, 'that these two objects have the same chemical composition ...'

I zoned out. I couldn't believe my luck. As it happened, that morning on the train from Cheshire I had been reading a crammer book about how to pass an Oxbridge interview, and that exact question had come up. The answer was all to do with molecular structure: in the case of the squash ball, the molecules were long, thin and springy, and loosely bundled together; in the case of the hard plastic, they were a rigid lattice.

I paused. If I were just to give the answer as I had read it that morning, wouldn't that be cheating? Outside, in the cold corridor, were half a dozen other hopeful eighteen-year-olds, each praying for their chance at academic glory. Didn't we all deserve a level playing field?

'I need to own up to something,' I heard myself saying. Silence. I blinked. 'We didn't do that at my school.' I looked Paul Raithby squarely in the eye. 'Though I am willing to have a crack at it ...'

A letter arrived a few weeks later, offering me a place. I was as excited as I'd ever been, and for several nights slept with it under my pillow. I had achieved a life ambition: I had won a place at Cambridge to study physics.

In actual fact I was mistaken; what I had in fact won was a place at Cambridge to study chemistry. For on my first day in harness, I realised that my ambitions didn't necessarily tie in with those of my new college, St Catharine's. I presented myself to John Shakeshaft and announced my intended speciality. 'Really?' he said, with some surprise. 'But you gave such an elegant answer to the chemistry question. I'm afraid Paul rather had you down as one of his!' I begged to differ. 'But you didn't even know what momentum was.' An off-day, I explained. I lived for Newtonian physics. 'Very well,' he said with a sigh. 'Welcome aboard.' He didn't look up until I was nearly at the door. 'Mr Miller?' I turned with great hope in my heart. Could this be the offer of a scholarship? 'What exactly *did* they teach you at your school?'

THE INTERESTING STUFF

Studying Natural Sciences was one of the hardest things I've ever done, and without any doubt one of the most rewarding. I was lucky enough to attend lectures by Stephen Hawking, Richard Feynman and Karl Popper, and to study alongside some of today's most prominent scientists. As far as the course went, we were thrown in right at the deep end, with Special Relativity, the theory where Einstein proposed that matter is a form of energy with his famous equation $E = mc^2$, which will come into play in the next chapter when we discuss the brand-new Large

Hadron Collider at CERN. We synthesised deadly poisons so toxic that one drop would be enough to kill the audience of a medium-sized fringe theatre. We learned how stars and galaxies were formed – more of that in Chapter 3 – and what causes volcanoes and earthquakes and global warming, which I shall treat you to in Chapter 7. It's this little treasure trove of knowledge that I want to share with you in this book; my own little mix-tape of scientific greats. It's an eclectic selection. There are some tub-thumpers in there, no question, as well as one or two forgotten album tracks that I always felt deserved greater attention – but, without exception, each and every one of the goodies I've included is a bona fide classic.

By the way, a quick word about footnotes. Now and then I just couldn't stop myself from adding a little bit more detail, and rather than break up the flow I've tended to put it in a footnote. Sometimes they act as a sandbox for me to do a little bit of maths; sometimes they are just a handy place to park an anecdote; hopefully the book reads as well with or without them, so please feel free to dip in and out of them as you wish. If this book is about anything, it's about enjoying science on your terms, purely for pleasure.

NOVEL QUANTUM EFFECTS

I don't want to give the impression that my entire time at university was spent working; that would have been a dreadful waste. I spent a very enjoyable year as my college's Entertainments

Officer, turned out for the Second XI football team and consumed an awful lot of what I can only describe as cooking lager. I was even instrumental in encouraging the student body to purchase a mirror-ball, which, I was very pleased to see at a recent reunion, still graces my old college's bar. Uncool as it may be, however, I adored my subject, and some of my happiest days were spent holed up in the library surrounded by textbooks, or in lectures hanging on the every word of the very men who had written them. I couldn't believe my luck. By the time it came to take my finals I was convinced my future lay in academic physics, and what better place to begin a PhD than the Cavendish Laboratory?

The PhD I signed up for was unfeasibly peachy; in fact, in 1988, it was one of the peachiest PhDs on offer. Professor Mike Pepper had founded the Semiconductor Physics Group ten years earlier, and it had grown to become one of the largest groups of its type in the world, employing some one hundred graduate students. It had something called a Molecular Beam Epitaxy Machine, which was able to make crystals of extremely high quality, and it had low-temperature fridges – huge great thermos flasks filled with liquid helium that were capable of reaching temperatures of a few thousandths of a degree above Absolute Zero.[1]

---

1 Temperature is one of those funny things that seems very straightforward at the everyday level, but when you get into it is really rather odd. Whereas there's no limit, as far as we know, to how hot something can be, there is a limit to how cold it can get – because, at some point, you will have removed every bit of heat it is possible to remove. This is the temperature we call Absolute Zero, and it crops up at about −273°C. That's an unsatisfactorily random number, but, of course, at the time that we invented the Celsius scale we were looking for something that conveniently described the temperature range of water; 0°C being its freezing point, and 100°C being its boiling point. Once you have to deal with larger temperature variations than that, you tend to use the Kelvin (K) scale, where Absolute Zero = 0 K, and 0°C is, you guessed it, 273 K. The surface temperature of the Sun, for example, is roughly 10,000 K.

So: semiconductor physics. Most people have a good idea of what a conductor and an insulator are, but what, I hear you ask, is a semiconductor? A conductor, after all, is something that an electrical current can easily flow along, like copper wire. An insulator is something that electricity won't flow through, such as plastic – which is why, incidentally, we use plastic to coat copper wire, safe in the knowledge that if we touch the plastic coating we won't get electrocuted.

A semiconductor, as you might have guessed, is something that conducts electricity a bit; not quite as well as a conductor, and not nearly as badly as an insulator. The best-known semiconductor is silicon – the main component of sand, glass and quartz – and it forms the basis of a huge variety of electronic devices. You might have heard of 'Silicon Valley', a nickname for the Santa Clara Valley area of San Francisco, coined because of the number of electronic-chip manufacturers based there. One popular alternative to silicon is gallium, and the group that I joined was in the habit of making very pure gallium crystals, adding different amounts of chemical impurity so as to vary how well they conducted electricity, and sandwiching them together.

Do this right, and you'd end up with a two-dimensional electron gas at the interface between the two sandwiched crystals, where, if you could get the temperature low enough, the mean free path of the electrons – that is, the average distance that they travel before banging into something – was very large, something like a thousandth of a millimetre. (I know that sounds ridiculously small, but the average distance travelled by an electron in a metal like copper would be about a hundred times shorter.) My PhD consisted of making tiny gold patterns on minuscule chips of gallium arsenide; when I charged up the gold patterns with a variable power supply, I would be able to create little patterns in the electron gas underneath.

The pattern that I was most interested in creating was a dot, the point being that weird things start to happen when you put very small things like electrons in very small boxes; in fact, you start to see a whole new landscape of behaviour that is nothing like the way that objects behave on the scale of everyday life. This is the strange world of quantum mechanics, of which we shall learn a lot more in the next chapter. Running an experiment was a lengthy process; cooling a chip down to near Absolute Zero could take the best part of a day and would quite often destroy my dot pattern in the process. Often the experiments would run through the night. I spent many thrillingly silent evenings camped out with just my sleeping bag and a pile of cheese sandwiches, watching and waiting, and – ever so occasionally – seeing quantum behaviour at first hand.[2]

ABSURD PERSONS PLURAL

Before we crack on with the briefest of descriptions of how this book will work, I do, of course, need to address one question that will no doubt have crossed your mind. How is it, you

2 The working title of my PhD (sadly, never completed) was 'Novel Quantum Effects in Quasi-Zero-Dimensional Electron Systems'. If you haven't guessed it by now, 'quasi-zero-dimensional' is just a fancy word for 'dot'.

may very well ask, that after studying so much science, for which I had so great a passion, I ended up in sketch comedy?

It's a hard thing to admit, but by the end of my first year as a graduate student I was beginning to recognise that, although I had a huge passion for the subject and a great deal of competency as an experimental scientist, I was never going to be in Stephen Hawking's research group. To put it another way, I was the physics equivalent of a session musician: technically able, perhaps, but it was never going to be my face on the album cover. At the undergraduate level it was fine just to be a fan of the subject – but, at the graduate level, it felt as if I needed to be leading the herd, not just following it. Somewhere, deep down, I couldn't help wondering whether I was on the right path. I'd never really tried my hand at anything else, and I was at a loss to know what to do.

Then, by chance, in the summer of 1989 the circus came to town.

The National Student Drama Festival was an annual event that toured the country, taking up residence at one or other of the British universities for a period of two years before moving on again. In 1989 it upped sticks and moved to Cambridge, and it just so happened that one of my friends at the time, Carole-Anne Upton, had landed the rather onerous task of trying to organise it.

Rather kindly, she offered me the opportunity to drive the judges around for the princely sum of £10 a day. Needless to say, I didn't have a lot of other offers and leaped at the chance. One of the perks of the job was being able to attend the workshops at the festival. The playwright Charlotte Keatley, author of *My Mother Said I Never Should*, was running one of the writing classes. Throughout the course of one very bizarre afternoon I found myself writing a sketch, having it performed by some actors and hearing an audience fall about laughing. It was a thrill like no other and I wanted in.

Unbeknown to me, Charlotte had taken up a residency at St John's College and had started a student writing group. When the festival ended, she invited me to join. Everyone else was writing rather serious plays about race and disability; I was the light relief, I think, and started to perform my sketches for the group. Slowly, I formed the idea that somehow, somewhere, there might be a future in it.

By the end of the following year I joined the Footlights, the university sketch society, as a writer, and was chosen to be one of the writer-performers in its centrepiece show, the summer tour. It was called *Absurd Persons Plural*, and during our run at the Arts Theatre in Cambridge, Griff Rhys Jones paid us a visit. He was incredibly enthusiastic and supportive, expressing an interest in buying some of the sketches from the show, one or two of which I had written. I started to get a trickle of work writing for *Smith and Jones* and even managed to wangle a bit part in one or two of the episodes.

A research group is a close-knit thing, and some of my extra-curricular activities were starting to raise a few eyebrows among my colleagues. This was in the early nineties, well before everyone had mobiles, and it was getting a bit embarrassing to be called to the department office with the news that it was my agent on the phone. When I landed a full-time job in Arthur Smith's comic play *Trench Kiss* alongside Caroline

Quentin, which was booked for a month-long run at the Edinburgh Fringe, I decided it was now or never. After all, the delights of science would always be available to me, whatever career I chose, since – as I hope this book shall prove – the joys of science are open to everyone. Comedy, on the other hand, might not come knocking twice.

Quitting a PhD is a big deal, as academic funding is unpredictable at the best of times and having a student leave before submitting doesn't look great on next year's grant application. Nevertheless, at twenty-five, I was already getting on a bit if I wanted to make it in the fickle world of showbiz, and my thesis was going to take at least another eighteen months to complete. I decided to bite the bullet and tell Professor Pepper the bad news: that I was going to leave the Semiconductor Physics Group and make a go of it in the world of comedy.

I'm not sure who was more relieved, him or me. 'Good for you,' he said, beaming like the Cheshire Cat. 'Do you ever watch that show, *Whose Line is it Anyway?* Chap who plays the piano, Richard Vranch, he was one of mine. Used to wear a dinner jacket to the lab. Really brightened the place up. If you bump into him, do give him my best.'[3]

If I was in any doubt about whether or not I had made the right decision, I soon got the confirmation I needed. That summer I directed the Footlights tour, *Cambridge Underground*, which, just like *Absurd Persons Plural* the previous year, played at the Cambridge Arts Theatre. Front row and centre for the opening night was my all-time hero, Professor Stephen

---

3 Richard Vranch, incidentally, unlike me, did complete his PhD. Show-off.

Hawking. He seemed to be enjoying the show immensely. I may not have been able to join him at the forefront of human knowledge, but at least I could give him a ruddy good laugh.

A LOVE AFFAIR WITH SCIENCE

I am a big fan of science, but I am the first to admit that there are vast swathes of it that do nothing for me whatsoever. What gets me going – and what I suspect gets you going too – is the big stuff: DNA, black holes, aliens and the end of the Universe. So this book is going to do something that books about science very rarely do: we are going to eat the pizza topping and leave the crust. By the crust, I mean ticker-tape timers (if you don't know, don't ask), osmosis (what is it with biology teachers and osmosis?) and anything to do with oxbow lakes (possibly just me). We are not going to fuss over the detail, except where the detail is delightful. We are going to talk broad strokes. We are going to get fancy. You and Science are going to blind yourselves to one another's shortcomings and have a wild, passionate affair.

As you begin this new exciting relationship, I know that you are, in some sense, damaged goods. Science has hurt you in the past. Not to begin with, of course. As a child, you and Science adored each other. After all, what toddler doesn't marvel at the Moon and the stars, and boldly thump whatever buttons it can get its hands on at the Science Museum? But as you grew older, your relationship became more difficult. Despite your efforts to build bridges, Science confused you, patronised you or, worst of all, bored you. Meanwhile, the arts — with their wafty libertine wiles — seduced you shamelessly. Possibly there was peer pressure, which, for teenagers, can be hard to ignore. Science, giggled your friends, is 'uncool'. You tried hard to resist. Maybe, for old times' sake, you made one last attempt to make it work — you did science A levels, perhaps — and for a while you rekindled some of the old magic. Sadly it was not to last; you lost touch after university and have since become little more than strangers.

Science, for you, is Unfinished Business.

Of course, your old flame has little direct relevance to you these days. There is probably not much call in your everyday life to go off hunting for the so-called God particle, otherwise known as the Higgs Boson. Whether there are many Universes or not, the kids still need taking to school and the bills have to be paid. Whether or not we are all falling slowly into a black hole (incidentally, we are) there are pictures that need hanging and parcels to be collected from the post office. But there's a niggle. Some part of you just can't forget how exciting the world of science once was, and can't help wondering if maybe, just maybe, it might be worth tracking your old friend down for one last ridiculous fling.

Well, I'm here to help you do just that.

I'm not for one second suggesting that you and Science should have made a go of it. I think the choices you have made — the responsibilities you have shouldered, the life you have built — are sacrosanct. I just think that the two of you could benefit from some, shall we say, easy company. And so — at the risk of stretching a metaphor gossamer thin – intellectually speaking, consider this book as a spare key to my place in town. It's opulently decorated, the fridge is well stocked, and I rarely use it during the week. You and Science are both welcome to drop by any time. If anyone asks, I shall swear blind that the two of you were never out of my sight, and we spent every evening *à trois*, watching subtitled Eastern European films – or reading Virgil in the original Latin: whichever feels the most plausible.

I see a flicker of disapproval in your eyes. Please don't think me improper. I don't want to break up your happy home. In fact, I want nothing to do with your home. I simply suggest that, once in a while, in order to appreciate it all the more, you open a skylight and look up at the stars.

FIRST PRINCIPLES AND LAST CHANCES

So let's get down to business. This is a book about science.

Quite possibly, the thought of that excites you; equally possibly, it makes you want to run a mile. Either way, let me reassure you: from here on in there is nothing but pure pleasure. Every idea within these pages is just as thrilling as the concept that we are all made from dead stars, and nothing in it is going to be any more difficult to understand. And none of it is going to be even remotely like hard work; rather than trudge wearily to the top of a mountain of scientific knowledge, we are going to parachute onto the summit and ski down.

In other words, what I'm promising is something slightly

different to the way these things are usually done. As you will know from the science you studied at school, and quite possibly from some of the popular science books you've had a spirited stab at, most science is taught according to a basic template known in the trade as First Principles.

The idea is to teach science from the ground up, starting with the basics (ticker-tape timers and graphs) and gradually adding complications until you get the full picture (Newton's laws of motion). It's the way that most of us were taught science at school and for the uninitiated it can often seem like some sort of cruel joke.

For example, take atomic theory. It went something like this: in our first year of secondary school, we were told that all stuff – trees and houses and small plastic models of David Tennant – was made up of ridiculously small particles called atoms and these were the smallest things you could divide matter into. The following year, we were told that actually there were plenty of things smaller than atoms, such as nuclei and electrons. The year after that, we were told that we had once again been lied to, and that nuclei were made up of protons and neutrons ... And so the whole sorry story continued, right the way through undergraduate physics and graduate physics, until you began to (a) lose faith that any scientists actually knew what they were talking about; and (b) wonder whether the entire profession were inveterate liars and con-men.

Using First Principles in a book like this is, in my view, massive overkill in terms of what a non-scientist needs to understand. After all, if someone is interested in Formula One, you send them to watch a race at Monte Carlo, not to the

University of North London to study the mechanics of the combustion engine. Those of us who wanted to be scientists studied the dry nuts-and-bolts stuff, the Newton and the uniform motion in a straight line, because we knew that the real treats (String Theory, multi-dimensional space, quarks and gluons) lay ahead – and, in order to be able to understand them properly, we needed the basics. If you want to build a Large Hadron Collider, you'd better hunker down and get a physics post-doc. If you want to gawp at one and imagine how cool it would be if it blew up . . . Well, you've come to the right place.

So that's the aim of this book: to throw you in at the deep end with lots of fully-fledged, fascinating science so you can learn on the job. Because it turns out that, although the thing that underpins all science – the maths – will be a closed book to us, lots of what scientists call the 'hand-waving' stuff isn't. After all, just because we don't speak the language doesn't mean we can't hang out for a couple of weeks on holiday and get by with hand signals.

What's more, as part of my covenant to you, I am going to give you the ultimate get-out-of-jail-free card. You don't need to understand anything. The purpose of other books may be to educate you, or provoke you, or challenge you; the purpose of this one is simply to entertain you. Relax. Breathe. There is no test. I want you to give yourself permission to have it all go right over your head, to grasp the odd fragment, to get the gist. If you find yourself reading a paragraph over and over again, unable to grasp it – and I'll be doing my level best to make sure that doesn't happen – then it's my fault, not yours; move on. This is not a science lesson. It's a science orgy.

CHAPTER 2

THE SUPER-DUPER
ATOM SMASHER

SIZE MATTERS

Have you ever wondered what the smallest thing in existence might be? Most of the objects we deal with in our everyday lives are, of course, round about the size of a human hand – and I'm not entirely sure that's a coincidence. The smallest division on the average ruler is a millimetre, largely for the reason that most people can't make out much that's a lot smaller than about a tenth of that. Imagine for a moment, however, that you are very into your arts and crafts, and you take up that peculiar hobby where you use a magnifying glass and a minute razor blade to carve a palm tree, say, or a smiley giraffe, into the end of a matchstick. For the end-of-year show, you want to pull something really special out of the bag.

What would be the smallest object you could ever hope to decorate?

The answer, my friends, is an electron. At least, it would be an electron if there was anything small enough to carve it with – which, of course, there isn't, because there's nothing smaller than an electron.[1] And the next biggest thing up from an electron? Now there, my friends, lies a story. Because the next biggest thing up from an electron is something utterly thrilling, called a quark.

Quarks – and 'quark', by the way, rhymes with 'squawk' – are fascinating things. So far, we have discovered six of them, but of these the only two you really need to worry about are the 'up quark' and the 'down quark', because they are what protons and neutrons are made of – and, as you know, protons and neutrons are what atomic nuclei are made of. All the things that you hold dear – your home, your family, your pet Schnauzer – all of them are just fancy ways of arranging quarks.

How do we know that quarks exist? Well, that, my friend, is where some very canny bits of kit called particle colliders come in. And the very latest in particle colliders is the world-famous Large Hadron Collider or LHC.

If its name sounds off-putting, just think of it as a kind of microscope: a window onto the bizarre world of the Very Small Indeed. With an optical microscope the best resolution we can hope for is around 100 billionths of a metre, small enough to see some of the larger viruses such as Ebola. An electron microscope is capable of making images of objects a

---

1 Not that we know of, anyway.

thousand times smaller than that, such as carbon atoms. But to investigate anything smaller than an atom – and a quark is nearly a billion times smaller than an atom – you need a different method altogether. And, crazy as it might sound, the method the LHC uses is to collide two protons together and examine what turns up in the wreckage . . .

A WHET OF YOUR APPETITE

I find it hard to put into words just how exciting I find the LHC. There's almost no way to over-exaggerate its significance for the future of physics and, by implication, the future of technology. Don't be put off by the fancy handle; hadron is just the family name for something made of quarks, and the whole thing could just have easily been called the Large Proton Collider, because that is its main function. It may be an extraordinarily complex bit of engineering, employing ten thousand of the planet's top scientists over a time-span of twenty years at a cost of some £4.4 billion – but at its heart is an extremely simple principle: get two protons moving in opposite directions at very high speeds, smash them into one another and look at what gets produced.

Why protons? Well, the answer is that there's no such thing, as far as we know, as a lone quark, so a proton is the next best thing. Protons are made of quarks, and if you make enough collisions, sooner or later one of the quarks in one proton will collide head-on with one of the quarks in the other proton and all sorts of interesting stuff turns up in the wreckage. Of the many things we might see, the most eagerly awaited is, of course, the Higgs Boson.[2] Again, don't be put off by the fancy name; the 'Higgs' bit is named after the Geordie mathematician, Peter Higgs, who first proposed the bit of maths that best describes it, and the 'boson' bit is another family name for the group of particles it belongs to. Why it is so important – so important it's been nick-named the 'God particle' – is something I shall take great delight in explaining in this chapter. Suffice it to say, if the Higgs is there, then the LHC is going to find it. Exactly how we will know we have found it is a fascinating specialist-knowledge secret that I shall be very pleased to let you in on.

As we shall see in the second half of the chapter, however, there's much more to the LHC than just the search for the Higgs. A whole new world is opening up for investigation, the world of objects as small as a tenth of a billionth of a billionth of a metre. Among the truly extraordinary things we might find are extra-spatial dimensions – yes, really! – and a whole host of new, as yet undiscovered particles. Not only that, but as we shall see, the LHC has been built to explore not just the next chapter of quantum physics, but to try to answer two of the most fundamental unanswered questions in cosmology too: first, if the Big Bang created equal amounts of matter and antimatter, where has all the antimatter gone? And second, if the Big Bang

2 Bosons carry forces and are often referred to as 'force particles'; they take their name from the Indian physicist who first described them, Satyendra Bose. The family name for matter particles is 'fermions', after the Italian physicist Enrico Fermi. More on him later. 'Boson' and 'fermion', by the way, are great words to bandy about at dinner parties.

sprayed matter out equally in all directions, as you'd expect, then how come that matter managed to clump together into stars and galaxies? We really need to answer those two questions, because they are absolutely fundamental to our understanding of how we came to be here in the first place.

Of course, one of the other talking points of the LHC has been the fact that the collisions are going to be of such high energy that, alongside a Higgs Boson, something else might get produced that is a lot less desirable: a black hole. That is a possibility, but before you throw this book away in a fit of nihilist angst and drain that raffia-covered bottle of dodgy Majorcan brandy that's been sitting in the cupboard ever since the late nineties, let me assure you that such a black hole is never, ever going to destroy the known Universe – exactly why will become clear very shortly.

THE ONLY GAME IN TOWN

I've only ever flown to Geneva for two reasons: one is to ski off-piste at Chamonix and the other is to visit CERN, the home of the Large Hadron Collider. I'd be hard pushed to say which one left me the more nervous.

CERN is, quite frankly, to physics what the Vatican is to Roman Catholicism. Established back in the 1950s as the Conseil Européen pour la Recherche Nucléaire, it was originally founded to explore the atomic nucleus but quickly switched to high-energy particle physics without ever bothering to update the acronym. The greatest scientific minds in the world are at work there: it was at CERN, for example, that Tim Berners-Lee created the World Wide Web as a way of sharing research information, thereby spawning the modern internet. Try that as an example the next time someone asks you what on earth is the point of pure academic research.

From Geneva airport, it's a thirty-minute cab ride to a rather unprepossessing huddle of what look like agricultural silage buildings on the suburban outskirts of a Swiss town called Meyrin. There is nothing whatsoever to tell you that you are approaching the nerve centre of the greatest scientific experiment ever attempted by humanity; every time I've been there it's been raining, and the nicest thing you could say about the place is that it always has an empty car park. Because the first thing to learn about the Large Hadron Collider is that nothing much of any importance is happening at street level; the exciting stuff – and, my god, it's exciting – is all happening underground.

What the LHC is, in essence, is one big circular underground racetrack for protons. And when I say big, I mean big: it's 27 kilometres in circumference. Buried about 100 metres below the surface, it is about the width of an average tube tunnel on the London Underground. Running down the middle of it are two pipes. One pipe has clumps of protons zipping along it in a clockwise direction, the other pipe has them zipping along it anticlockwise. At four points on the circumference, the two pipes cross. These four crossing points are where the protons can be made to collide, and huge detectors have been built around them, which are capable of sifting through the debris of the collisions and identifying all the different particles that have been produced. So, if you like, the LHC isn't really one experiment, but four; each detector has been designed by a different research team with their own sets of goals and priorities. Of these four experiments, two of them – ATLAS and CMS – are out-and-out rivals, both of them looking for the Higgs. The other two – ALICE and LHCb – are looking for clues as to what happened in the very early Universe, just fractions of an instant after the Big Bang. They hope to throw some light on two of the most troublesome problems in modern physics: the scarcity of antimatter and the 'lumpiness' of the present-day Universe.

THE SEARCH FOR THE HIGGS

So what's the big deal about the Higgs particle? To understand that, we need to talk a little bit about what physicists call the Standard Model of particle physics – and to understand the Standard Model, we need to grasp a couple of things about particles and forces.

You might remember from your physics classes in school that there are four types of forces in nature. Three of those four forces are of roughly the same strength: the electromagnetic force, which operates between particles that have electric charge; the strong force, which operates between quarks; and the weak force, which enables radioactivity. The other force, gravitation, is much weaker than the other three, and really comes into its own for large objects like stars and planets. The gravitation of the Earth, of course, is what's holding you in your seat right now and stopping you spinning off into outer space.

You might also remember from your schooldays that the standard explanation of how forces worked was that particles produced fields, which then acted on other particles. For example, an electric charge produced an electric field; when another electric charge was placed in this field, it experienced a force. I, for one, always found this explanation a little unsatisfactory: how did the second charge 'know' that there was an electric field present? If that ever bothered you too, then fret no more; quantum physics has a much better explanation. The answer is that every field has a carrier particle, or boson – as in the Higgs Boson. The electromagnetic field between two electrons, for example, is communicated by a particle of light, otherwise known as a photon.

Remember that Mr Bailey said mathematicians are lazy? Well, theoretical physicists are mathematicians, and they are far too lazy to be dealing with four different forces – so one of the chief goals of physics over the last fifty years has been to try to simplify this picture. The gut feeling is that nature just can't be this complicated, and these four forces are all related to one another in some fundamental way. In a nutshell, we're about three-quarters of the way there, and the theory that we have is called the Standard Model.

Things took a big leap forward in the early sixties, when a threesome called Glashow, Weinberg and Salaam showed that the electromagnetic force and the weak force were one and the same. They called this new unified force the electroweak force; to make the maths work, however, they proposed something rather radical. They begged to suggest that all fundamental particles had no intrinsic mass. That's right: no mass. Instead, they said, there's another undiscovered field out there in the Universe and, like all good fields, it has a carrier particle. Some fundamental particles, like the photon, zip across this field with scarcely a by-your-leave, and they therefore appear massless. Other particles, so the theory goes, aren't nearly so lucky and get dragged down so much by the field that they appear to have a lot of mass. This new field borrowed a bit of maths from a British chap called Peter Higgs, and was called the Higgs field in his honour. And the carrier particle that communicates this field quickly took the name of the Higgs Boson.

After an enormous group effort in the early seventies, the strong force was also combined with Glashow, Weinberg and Salaam's electroweak force and the resulting mash-up became known, quite simply, as the Standard Model. Some of the ideas at the heart of it are gloriously off the wall, but it really does seem to work. In fact, it's not overdoing it to say that the Standard Model is one of the most successful theories ever invented.

A really good test of any scientific theory is its success in predicting things that no one could have guessed at, and doing so with great accuracy. To give you a taste of just how impressive the Standard Model is, all you need to know is that it proposed that the weak force would have a carrier particle, just like the electric force. In fact, it proposed that it would have two, the W and the Z (W for weak, and Z for no good reason that I can work out). It also made a firm prediction of their masses: 86 proton masses for the W and 98 proton masses for the Z. This was back in the late sixties, in the days when particle accelerators were capable of collisions only equating to ten proton masses. Well over a decade later, in 1981, at CERN, a forerunner of the LHC called the Super Proton Synchrotron – a synchrotron, by the way, is just a fancy name for something that accelerates charged particles in a circle – finally reached the energies required. The W and Z particles were right there, as predicted, and their masses were exactly the same as predicted by the Standard Model.

THE GHOST IN THE MACHINE

So to sum up, the Standard Model is an extremely impressive piece of jiggery-pokery, which manages to unify electromagnetism with the weak and strong forces and has been one of the most successful scientific theories ever created. To make it work, however, physicists have had to propose the existence of a new particle, the Higgs Boson, which is responsible for giving the other fundamental particles their mass. Finding the Higgs is important, because if it exists, it confirms the Standard Model. Unfortunately, so far, the Higgs has failed to turn up to the party.

So why has no one ever found a Higgs? Well, one possible answer – and at this stage we can't rule it out – is that it quite simply doesn't exist. The Standard Model might be wrong in some important respect that we just don't know about yet.

Another possibility is that, until now, we haven't had collisions of a high enough energy to be able to produce it. To clarify what I mean by that, let's talk a little bit about the weirdness of the quantum world and exactly what is going on inside a proton collider like the LHC.

In some respects, the collisions that happen in the enormous detectors over in CERN are just like the collisions you might see on the average snooker table. Energy, for example, is conserved in knockabouts between protons in the LHC just as energy is conserved in the trick shots down at the Pig and Whistle. The difference is that, on a snooker table, you'd expect to have the same number of snooker balls after any given collision as you had before it. Well, in the world of the extremely small – or the quantum world, as physicists like to call it – life isn't quite like that. When you collide fundamental particles together, you sometimes end up with a greater number than you started with.

You probably already have a feel for why, whether you realise it or not, because the answer is contained in the most famous equation in physics: $E = mc^2$. Matter, as this equation so neatly expresses, is a form of energy. An atomic bomb is a sobering demonstration of just how much energy a little bit of matter can release, once you know how to unlock it – round about a kilogram of plutonium fissioned in the atomic bomb that destroyed Nagasaki, or about the weight of one bag of sugar.

Whereas an atomic bomb takes matter and converts it to energy, one of the main tricks in a particle collider like the LHC is to work the process the other way around; in other words, to convert energy into matter. Put simply, you get the particles involved in the collision up to really high speeds, so that they have lots of energy, then cause collisions that use this energy to create new particles. The higher the energy of the collision, the bigger the new particles you can potentially produce. With a decent-sized synchrotron, you can accelerate protons to quite extraordinarily high speeds; in the case of the LHC, they can reach 99.9999991 per cent of the speed of light, at which point they whizz round so fast that they complete a full 27-kilometre lap 11,000 times a second. At these ridiculously high speeds, the energy of each proton is equivalent to about 7,500 times its rest mass energy.[3] That's a lot of energy knocking around, ready and willing to be converted into new, as yet undiscovered particles.

It's this energy factor that may explain why we have never seen a Higgs in any of our previous colliders; the collisions that were taking place were quite simply not of a high enough energy. The forerunner to the LHC, the Large Electron–Positron Collider or LEP, was able to rule out a Higgs Boson mass of less than 122 times the rest mass of a proton, and at the time of writing the data seem to be homing in on a Higgs that weighs in at around 125 proton masses. Since the LHC can potentially produce particles

---

3 Once you start getting up to a decent fraction of the speed of light, your mass starts to noticeably increase. This is also expressed in Einstein's famous equation, since the '*m*' bit actually stands for:

$$m_0 \times \frac{1}{\sqrt{1-\left(\frac{v}{c}\right)^2}}$$

where m_0 is your mass when you're not moving, v is the speed you are travelling, and c is the speed of light. As v gets closer to c, in other words as you get closer to the speed of light,

$$\sqrt{1-\left(\frac{v}{c}\right)^2}$$

gets smaller and smaller, and m – the total mass – therefore gets bigger and bigger. I'll be mentioning this again when we get to the subject of space travel in Chapter 8.

that are up to a thousand times heavier than a proton, we can be fairly confident that, if the Higgs exists, the LHC will find it.

And what will a Higgs look like? The truth is, we won't observe one directly; there's no detector for a Higgs. As with Santa Claus, we'll have to infer its presence by the evidence it leaves behind. In the case of a Higgs, of course, that won't be a half-eaten mince pie and an empty brandy glass, but rather a tell-tale, Higgs-sized, missing bit of energy. Millions of interactions will take place in the detectors of the LHC during any given experimental run, and computers will sift the data from these umpteen events, looking for anything out of the ordinary. The production of a Higgs will leave a certain 'signature'; any events that mimic that signature will be analysed in great detail to determine whether they unambiguously imply that a Higgs has, indeed, been present. And what, you might ask, if the detector is faulty? What if all this complicated machinery goes awry and creates a phantom Higgs signal? Well, that, my friend, is why there are two experiments at the LHC that are on the lookout for this most elusive particle: ATLAS and CMS. The idea here, of course, is simple: if ATLAS says they have found a Higgs, you can be very sure that their competitors at CMS will not be satisfied until they find one too.

BLACK HOLES

Before we move on to talk about the other gut-wrenchingly exciting things that could conceivably show up in the detectors of the LHC, like extra-spatial dimensions and supersymmetric particles, let's take a brief but enjoyable diversion into the subject of black holes.

You'll recall that, back in early September 2008, there were two races going on: one at CERN, where physicists were working round the clock to get the LHC ready for switch-on; and another in the Hawaii District Court where one Walter L. Wagner and his associate Luis Sancho were doing their damnedest to get an injunction to try to stop them. The Hawaiian suit claimed that the collisions in the LHC could create – among other scary things like magnetic monopoles and strangelets[4] – microscopic black holes that would devour the planet.

As the day of reckoning drew ever closer, there were ever more stories in the press, articulating a mounting concern that switching on the LHC was going to cause a black hole, leading to the end of, if not the known Universe, then certainly some obscure parts of Switzerland. As a matter of fact, it became pretty much impossible to talk about the LHC in everyday conversation without someone cracking some sort of black-hole gag, rather in the same way that it has recently become practically illegal to venture out on a rainy day without an ironic reference to global warming.

---

4 A magnetic monopole, by the way, is a particle that is a magnet with only one pole. The jury's out on whether they exist or not; they have never been observed, but, as with the Higgs, that may be because they are too large to have been created in any pre-existing particle collider. A strangelet is a hypothetical particle that contains strange quarks, but remains stable. All known particles that contain strange quarks, like the K-meson for example, are highly radioactive and decay pretty sharpish into smaller particles made of common-or-garden, up and down quarks, so this would be a bit of a turn-up for the books.

Never mind that switching on the LHC wasn't even going to involve any collisions, just the two proton beams circulating in the main ring; the world was now expecting something spectacular, and anything less than some sort of latter-day Big Bang was going to feel like an enormous anticlimax. So when, a few days after switch-on, there was indeed a small bang of sorts when one of the superconducting magnets quenched after a liquid helium leak, the world's press had the sitcom-style, act-break resolution that it so clearly craved. The subtext of much of what was written was clear: if these numpties can't even solder a magnet together properly, then why on earth are we allowing them to run an experiment that could – this is a black hole we are talking about, after all – suck us all to kingdom come?

Build a piece of technology as culturally significant as the LHC and you are always going to attract a few nutters, but for me the real irony of the whole black-hole debacle is that the real science, as ever, is even more mind-boggling than the pseudoscience that calls it into question.

To get straight to the point, we know the LHC can't destroy the planet because of the existence of something truly bizarre called cosmic rays. And for sheer weirdness, give me cosmic rays over black holes any day.

THE ABSOLUTELY FRIGGING GINORMOUS PROTON COLLIDER

Cosmic rays are big news. The name is a bit misleading as they aren't really 'rays' in the *Star Trek* sense, but particles. Most of them – round about 90 per cent – are protons, just like the ones in the LHC. The rest are the nuclei of heavier atoms, such as iron. The LHC, as you know, can accelerate a proton up to energies equivalent to around ten thousand times its rest mass; the fastest-moving cosmic rays have energies of about ten thousand *million* proton rest masses.

So where do they come from? The truth is, we can't say for sure. They bombard the Earth from every conceivable direction and with a vast range of energies. Some of the lower-energy cosmic rays appear to originate in solar flares on the surface of the Sun. Those of middling energy are likely to be coming from within the Milky Way galaxy; supernovae, like the one I mentioned at the beginning of the book, are one possible candidate. The highest-energy ones are assumed to be coming from outside the galaxy, and there is no known process that could have created them. It is as if, somewhere out there in the furthest reaches of the Universe, there was the most colossal particle accelerator in Creation. And whenever you have a fast-moving particle, you are asking for a collision.

On their arrival in the upper atmosphere, cosmic rays collide with the molecules of the air, producing showers of new particles in exactly the same way as proton–proton collisions do in the LHC. If you like, you can think of the Earth as being one huge collision experiment that has been running for roughly 4.6 billion years. The very first discoveries in particle physics were made not in man-made accelerators, but in so-called 'cloud chambers' at the tops of mountains where the products of cosmic ray collisions were plentiful.[5] The strange quark, for example, was first encountered in the form of a particle called a K-meson (a meson is a particle that contains a quark and an antiquark) in exactly this way. The LHC is far from an abomination against nature. Quite the opposite – it is a pale imitation of one of the most fundamental processes in the cosmos.

So when some wonk in Hawaii thinks that the relatively puny collisions of the LHC might cause a microscopic black hole that will devour the planet, or that some other scary hypothetical thing like a strangelet or a magnetic monopole might suddenly leap into existence and start munching its way through Geneva airport, we can be sure he doesn't really know his physics. Every object in the Universe you can possibly think of – neutron stars, white dwarfs, black holes, galaxies, stars, planets, footballers, footballs, air molecules, protons, electrons – have been bombarded with high-energy particles since the beginning of time, and they are all still here. And if high-energy particle collisions are capable of creating weird secondary particles like miniature black holes, or strangelets, or magnetic monopoles, or diddly-dong-whatsit-MacTavishes, then every object in the Universe has also been bombarded with those since the beginning of time too. And guess what? Every object in the Universe is still here! To summarise: (i) the Universe exists; (ii) the Universe is constantly bombarded with every particle in Creation. It is therefore

---

5 A cloud chamber is basically a box full of cold vapour. When a charged particle like a cosmic ray passes through it, the vapour condenses along the line of its track. If you are so inclined, you can then take a photograph and examine the particle tracks at your leisure. If you take the added precaution of applying a magnetic field of known strength, you can work out the mass of the particle from the curvature of the track in the magnetic field. Clever, eh?

a racing certainty that (iii) no particle in Creation is capable of destroying the Universe. Not a certainty; science never provides us with those. But a probability so close to a certainty that no sane person would ever bet against it.

PSEUDOSCIENTISTS' CORNER

The scientists at CERN are far too polite to say it, but if you ask me, what the LHC-black-hole controversy boils down to is the difference between science and pseudoscience.[6] One of the main problems we have in our culture is that the media, in general, can't tell the difference. In some ways, that's under-standable; after all, pseudoscience, by definition, makes its way in the world by mimicking science and sometimes manages to do quite a good job. And when you take into consideration the fact that most people who work in the media have arts degrees and are positively phobic about science, you can see where there's a weakness to be exploited.

It's hard to imagine a news editor of a national newspaper who wouldn't know of the existence of Shakespeare, but all too easy to conjure one that has never heard of cosmic rays.

---

6 If extraordinarily patient explanations are your thing, check out the exhaustive CERN paper, 'Astrophysical implications of hypothetical stable TeV-scale black holes'. It runs to around 100 pages. As a watertight refutation of the proposition that a microscopic black hole might destroy the planet, it is undoubtedly impressive, but reading it can't help but leave me, for one, with the nagging question: Is this the best way for some of the most talented particle physicists in the world to be spending their time?

This same news editor might have a healthy scepticism about, for example, errant pseudoscientific hogwash like homeopathy, but how sure can we be that they would know their onions when it came to particle physics? Particularly when, as in the case of the Hawaiian injunction, the petitioners are able to muster just enough technical terms – black hole, strangelet, magnetic monopole – to sound plausible?

The problem, of course, is that the more like a scientist a pseudoscientist appears, the more dangerous the nonsense they peddle. It's one thing to dress up in a black silk cloak and a pointy hat with stars on it and claim to be able to tell the future; it's quite another to describe yourself as a 'nuclear physicist', as one of the Hawaiian petitioners did, when you have absolutely no recognised qualification in the field. No one takes the horoscopes in newspapers seriously – at least, I hope they don't – but several people took the threat of cataclysmic black holes seriously enough to make death threats to scientists on the LHC project. To paraphrase John Ellis, one of the greatest particle physicists of his generation and a leading light of the LHC Safety Assessment Group: 'The public may be safe from the LHC, but is the LHC safe from the public?'

THE WEIRD AND THE WONDERFUL

Notwithstanding crackpot lawsuits and the febrile imaginations of self-proclaimed 'nuclear physicists', the sober fact remains: while it's true to say that no one is holding their breath while they wait for a magnetic monopole or a strangelet to show up in the detectors at CERN, many genuine scientists who do study the very real subject of particle physics do believe that the LHC might be capable of producing a microscopic black hole. And that, my friends, is something to be celebrated, not lamented. So what exactly is a black hole, and how would a microscopic one show up in the LHC's detectors?

There's nothing that complicated, really, about the idea of a black hole; it's just a very dense object that has completely collapsed under its own weight. Density, of course, means the amount of matter packed into a given volume of space. An everyday-sized lump of ordinary matter, like, say, a piano, isn't dense enough to collapse under its own weight – any tendency to do so is easily balanced by the forces between its constituent atoms. But stars are much denser than ordinary matter; make one big enough and things start to get interesting. Take our own Sun, for example. The inward gravitational pull on the matter at the surface of the Sun is enormous; the only reason the Sun doesn't collapse is because of the outward pressure produced by the nuclear reactions at its core. But what's going to happen to the Sun when those nuclear reactions eventually burn themselves out?

The answer is that the Sun will partially collapse; in fact, it's expected to be squashed to the size of the Earth, about a hundredth of its present size. Looking around the Universe, we can see that many stars are several hundred times the mass of the Sun – so what happens when they burn out? A black hole, that's what. Unable to support their own gargantuan weight, they will be compressed so small that they will be no bigger than an electron. If you are far enough away from a black hole, its effect is just the same as that of any other large gravitational body, like a star. But get close enough and you will be sucked inside, never to return. That, of course, is why they get the name 'black hole'. Once close enough, not even light can escape.

As we shall see in the next chapter, 'We Are Slowly Falling into an Enormous Black Hole', we believe that most galaxies, including the one we call home, the Milky Way, have a black hole at their centre. The one at the centre of our galaxy weighs in at an astonishing 4.3 million solar masses, but an object of any mass can become a black hole if it's dense enough. To put it another way, big stars form black holes because they are dense, not because they are big or because they are stars. You yourself could become a black hole if I could squash you into a small enough volume. And where else do you get a lot of matter in a tiny volume of space but in the collisions taking place in the LHC?

The energy of a proton in the LHC is around seven thousand proton masses; a collision between two protons yields twice that much energy, around fourteen thousand proton masses. That's a lot of energy packed into a very small amount of space. The big question is: is it dense enough to cause a microscopic black hole?

The routine answer, unfortunately, is no, and not by a very long chalk. If gravity is as weak as it appears to be in everyday life, the collisions between protons in the LHC simply aren't of a high enough energy to cause a black hole; in fact, they are about a thousand million million times too small. But there are some theories around – String Theory being one of them – that suggest that quantum-scale gravity might be stronger than everyday-scale gravity. And if that turns out to be the case, quantum-sized black holes might be a very real possibility.

If one does show up, however, all the theoretical predictions are that it won't hang about. One of the leading lights of the maths of black holes is Stephen Hawking, who has predicted that black holes evaporate due to a process called 'Hawking radiation'. The smaller the black hole, the quicker the burn-off rate; in the case of a microscopic black hole in the LHC, the evaporation would be pretty much instantaneous and would appear as a sudden splurge of particles of all types in all directions. So, if you like, a quantum-sized black hole isn't really black at all; it's more like a miniature Sun.

It's almost not healthy to get too excited about all of this, because String Theory is far from proven and LHC-created black holes might so easily be complete pie in the sky. Suffice it to say, if they are created – as a tool for understanding gravity, extra dimensions, the Higgs and anything else you care to mention – quantum black holes would be second to none. Far from dreading their arrival, I am hanging out the bunting, praying that we are lucky enough for one to pay us a visit.

THE FINAL FRONTIER

Earlier in this chapter, I talked about how the story of particle physics to date is really the story of the unification of the four forces of nature: the strong force that holds nuclei together; the weak force that causes radioactive decay; the electromagnetic force that acts on electrical charges; and finally, gravitation, which is so much weaker than the other three that you notice its cumulative effects only in large lumps of matter, such as an apple in the gravitational field of the Earth. Our best efforts so far have yielded us the Standard Model, which combines the strong, weak and electromagnetic forces at the expense of creating a new field, the Higgs field, which is responsible for giving elementary particles their mass.

It goes without saying that particle physicists haven't decided that's close enough for jazz and called it a day. Instead, ever since the sixties when the Standard Model emerged, they have been grafting like stink to come up with a new theory that incorporates gravity into this picture. This mathematical Holy Grail is called the theory of everything, or TOE for short.[7] And coming up with something that does the job has been a bit of a struggle.

So far it's been a bit of a two-horse race. As a very broad generalisation, the theorists have divided themselves into two camps, and it has to be said there is not a lot of love lost between them. One camp, the Gauge Theorists, have pressed on with the same maths that yielded the Standard Model. The other camp – and it has to be said they're younger, trendier and have had slightly better PR – are called the String Theorists. The maths they use is a clean break from Gauge Theory, and involves lots of fun new ideas like strings, higher dimensions, supersymmetric particles and microscopic black holes. One or more of these bad boys may show up in the ATLAS and CMS

---

7 The name, by the way, was coined by the same John Ellis who sits on the CERN Safety Assessment Committee.

particle detectors, which have been designed to find the Higgs, though I think it's fair to say it's a bit of a long shot. So let's have a quick shufti at the theory and the sort of experimental evidence that the scientists at the LHC will be looking out for.

STRING THEORY AND THE LHC

As is probably becoming evident, there really seems to be no limit to the number of particles, both fundamental and composite, that inhabit the quantum world. Gone are the days of the early twentieth century, when nature simply served up electrons, protons and neutrons; ever since the invention of the first cloud chamber particle detectors, stuff has been turning up that no one wanted or needed, and physics has been fighting a running battle ever since to fit it all into some sort of coherent picture.

The first of these unexpected visitors was a particle we now call the muon, a negatively charged particle like an electron but 200 times heavier. 'Who ordered that?' quipped Nobel Laureate physicist Isodor Rabi. This was in 1936; by 1955, when the kaon and the pion had both put in an appearance, everyone was starting to wonder where all this was leading.[8] In his Nobel acceptance speech of that year, the American Willis Lamb, recognised for his work in refining our understanding of the

---

8 A pion, like a kaon, is a particle that contains a quark and an antiquark. Its existence had been predicted twenty years earlier by the Japanese physicist Hideki Yukawa and earned him a Nobel Prize. The pion is the boson that carries the strong force between neutrons and protons in atomic nuclei. The gluon also carries the strong force, but between individual quarks, such as those within a proton or neutron.

electron, made one of physics' finest jokes when he said, 'The finder of a new elementary particle used to be rewarded by a Nobel Prize, but such a discovery ought now to be punished by a 10,000 dollar fine.'

So, just for fun, let's have a brief saunter around the particle zoo. Don't worry about the details; the point is just to give you an idea of the John Lewis-like range available. First there are the so-called leptons (from the Greek, meaning light) consisting of the electron, the muon, the taon and their respective neutrinos. All of these have antiparticles. Then there are the six quarks: up, down, top, bottom, charm and strange. All of these, too, have an antiparticle twin. Then there are the hadrons, made up of groups of quarks: mesons, which contain two quarks, such as the pion, kaon, and J/Psi; and baryons, which have three, such as our favourite proton and neutron – as well as more exotic creatures such as the Sigma (two up quarks and one strange quark) and the charmed Sigma (two down quarks and one charm quark). Confused? Oh, wait – I forgot the gauge bosons of the electromagnetic, strong and weak forces: the photon, the gluon, the W and the Z.

It's a testament to the power of Gauge Theory that, in its capable hands, this complicated picture has been resolved into a very simple one: matter particles and force particles. As we've already seen, though, there's a problem: it leaves out gravity. If gravity worked along the same lines as the other forces, we'd expect it to be carried by its own gauge boson, the graviton; so far, however, all attempts to make the maths work have proved fruitless, and nothing resembling a graviton has ever cropped up in a particle detector.

String Theory, on the other hand, takes a very different approach. Rather than see matter particles and force particles as two distinct and separate entities, String Theory takes what my mother would call a 'same horse, different jockey' approach. The gist of it is that force particles and matter particles are simply different modes of vibration of much more fundamental objects called strings. These obviously aren't strings in the normal sense, but mathematical objects, like quarks, that might never be directly observable.

An interesting side effect of this approach is that, rather in the same way that Gauge Theory had to invent a Higgs Boson to make the maths work, so String Theory also needed to invent some stuff to help the equations make sense: higher spatial dimensions and supersymmetric particles. Higher spatial dimensions means a number greater than the four we are used to dealing in;[9] in fact, the most common branches of String Theory deal in ten dimensions. I know what you're thinking: if there were more than three spatial dimensions, you'd have noticed them. But that's not necessarily the case; what if they were extremely small and inaccessible to you? The surface of a lake might look smooth, but get close enough and you can see that there are eddies and ripples and all sorts of texture not immediately apparent on a greater scale. What looks two dimensional on the human scale – the flat surface of a lake – is a completely different, three-dimensional matter to a little bug that lives on the surface of the water. So too might three

---

9 Three dimensions of space and one of time – or, as we shall see in the next chapter, four of spacetime.

smooth spatial dimensions look like four or more dimensions to a creature that was small enough to see them. Indeed, some of these additional spatial dimensions might be accessible to the particles created in the aftermath of collisions in the LHC – in which case, like the Higgs, they will show up as missing bits of energy. If they do, you can guarantee that String Theorists, who have so far lacked an experimental test of their work, will be on it like a ton of bricks.

One of the consequences of higher spatial dimensions, which I hinted at earlier, is that they might provide an explanation of why gravity is so weak when compared with the other three forces. Fasten your seat belts because, typically for String Theory, it's a bit of a weird one. The idea is that the four-dimensional space that you and I love isn't the only space that's accessible to gravity; perhaps, unlike the other three forces that are trapped in our world, gravity has the freedom to leak over into the other spatial dimensions – and, therefore, its strength gets diluted. No? Well, I told you it was weird.

Supersymmetry, on the other hand, is the straightforward idea that every particle we know and love has a heavier twin. Again, making this assumption helps the maths of String Theory, and the search for supersymmetric particles will be a fascinating aspect of the LHC project. There is a major riddle in cosmological physics, in that we are not at all sure what most of the Universe is made of. We call the mysterious stuff 'dark matter' and supersymmetric particles are one possible candidate. If we can create them in the collisions in the LHC, we will be able to rule them in or out of the whodunnit.

The great advantage of String Theory – and one of the

reasons it has been so successful despite being a bit 'out there' – is that it provides a model that incorporates all four forces, including gravity, into one picture. In the absence of any hard evidence to support it, it has become something of a bone of contention in the physics community. The LHC is extremely exciting because it may well provide the first indications of whether Gauge Theory or String Theory is on the right track. The Higgs, if discovered, will provide a huge boost to the Gauge Theorists' Standard Model. Extra-spatial dimensions, or supersymmetric particles, or – even more tantalisingly – a quantum black hole, would be an enormous boost to the String Theorists. Or quite possibly, both theories will be proved wanting, and we shall have to invent some entirely new physics. In some ways, that would be the most interesting result of all.

ANTIMATTER AND THE BIG BANG

As we've seen, two of the experiments taking place at the LHC involve enormous, general-purpose detectors called ATLAS and CMS, which are capable of tracking down anything from a Higgs particle to an extra-spatial dimension. In this respect, the LHC is like an incredibly powerful microscope, helping us to see the very finest details of the quantum world. But as I've already mentioned, there are two more experiments taking place at the LHC, called LHCb and ALICE, which in many ways are more like incredibly powerful telescopes, transporting us back to a few fractions of a second after the Big Bang. One of them, LHCb, deals with antimatter. The other, ALICE, is attempting to re-create a phase of matter that hasn't existed since the Universe was in short trousers.

ANGELS AND DEMONS

Antimatter is amazing stuff. It turns out that every fundamental particle we have so far discovered has an antimatter partner, with the exact same mass but with the opposite charge. The antiparticle of the electron, for example, is the positron. The kicker is that, if a particle and its antiparticle meet, they annihilate one another, producing photons.

The particle collider that preceded the LHC, the Large Electron–Positron Collider, used collisions between very high energy electrons and positrons to create new particles and was capable of reaching a total collision energy of 200-odd proton masses.[10] In fact, there was a great drama towards the end of the millennium when many physicists working on the LEP believed that they were seeing evidence of a Higgs at around 122 proton

---

10 There are great advantages, incidentally, in using electron/positron collisions to create new particles; in a sense they are much 'cleaner' because the particles involved simply annihilate one another and produce a shower of new particles. When you collide protons, on the other hand, what you are really trying to do is get one of the quarks in one proton to collide with one of the quarks in the other proton. Needless to say, that's not easy to do, and means you have to create a heck of a lot of collisions, as well as do a lot more computing of the data, first to sift out the near-misses from the quark-on-quark action, and second to work out what the hell's going on. The problem with accelerating electrons and positrons in a circular synchrotron like the one at CERN is that you can't get their energies much above 100 proton masses, because they start giving off enormous amounts of radiation. Hence, the switch to protons – and a much bigger headache in analysing the results.

masses; bureaucracy prevailed, however, and the LEP was dismantled to make way for the LHC. It was a tough call, but turned out to be the right one, as it later became apparent that the signal was not as reliable as they had first thought. In the end, 122 proton masses was set as a lower limit on the mass of the Higgs.

Have no doubt about it, antimatter is very real, and at the heart of one of the main mysteries in physics: namely, why is there so little of it found in nature? We believe that equal amounts of matter and antimatter were created in the Big Bang, so how can it be that, when we look around the Universe, antimatter appears to be in such short supply?

In one sense, of course, this is very good news. If all the quarks and antiquarks created in the Big Bang had instantly annihilated one another, there would have been nothing left except photons. There would be no galaxies, no stars, no planets and no physicists to wonder where it all came from. It was a close-run thing: most of the quarks and antiquarks created in the Big Bang did annihilate, and the photons left over can still be seen today in radio telescopes all around the world. For every billion quarks that annihilated with an antiquark, however, one quark survived unscathed and helped to make up the matter that we see around us in the Universe today. That's a tiny difference, but so far we haven't found anything in the laws of physics that gets anywhere near explaining it.

One possible culprit is the weak force, which does appear to behave ever so slightly differently with matter and antimatter. The differences we've seen so far don't even come close to explaining a one-in-a-billion survival rate, but they are a vital clue in the search. The LHCb detector – b stands for bottom, as in bottom quark – has been built to measure the decay rates, due to the weak force, of the B-meson and its antiparticle, the anti-B-meson.[11] If the weak force treated matter and anti-matter in the same way, you'd expect the decay rates to be identical, but they aren't; the B-mesons decay ever so slightly slower than their antiparticles. Investigating this difference in decay rates at the never-seen-before high energies of the LHC is an important goal of LHCb, and will hopefully teach us more about the weak force and its role at the dawn of Creation.

THE LUMPINESS OF CREATION

The remaining experiment of the LHC is the quite extraordinary ALICE. The name stands for A Large Ion Collider Experiment – but don't hold that against it, because what's going to be happening within it can't be diminished by a rubbish acronym. Unlike ATLAS, CMS and LHCb, which are all designed to study proton–proton collisions, ALICE is going to do something quite different: it is going to take the largest ions it can get its hands on – lead ions, in fact – accelerate them to relativistic speeds and then smash them together to try to re-create a state of matter that last existed just fractions of a second after the Big Bang.

---

11 Now this is getting a bit complicated, but since you might be interested: the B-meson is made of a bottom quark and an up or down antiquark. An anti-B-meson consists of a bottom antiquark and an up or down quark. At this point, your brain starts to fry and it becomes easier to use a notation. Particle physicists therefore use the handy symbols u, d, s, c, t , and b to stand for the up, down, strange, charm, truth and beauty quarks. To denote an antiquark, you just put a bar over the top, as in $\bar{u}$. A B-meson is then $b\bar{u}$, an anti-B-meson is $\bar{b}u$. Maths, you see? It's for lazy people.

Back in the day before quarks and gluons began to solidify into the baryons we know and love like protons, neutrons and mesons, they existed in a gas-like state that physicists call a quark–gluon plasma. In the aftermath of a high-energy lead ion collision, CERN's physicists calculate that a blob of this stuff will be formed with a temperature some 100,000 times that of the Sun. Studying such a blob may shed light on all sorts of questions in particle physics, but it has all sorts of implications for cosmological physics too, since in many ways it will have the properties of a baby Universe.

One of the burning questions of cosmology is how it came to be that matter clustered together to form large structures like stars and galaxies. All things being equal, you'd expect that, after the Big Bang, matter would just spread out evenly, with every particle becoming more isolated from the rest as the Universe expanded. Needless to say, that didn't happen: the baby Universe must have had lumps in it, and the secret of what caused those lumps may very well lie in the quark–gluon plasma of the ALICE experiment.

ONE SMALL STEP

I was three when the *Eagle*, the lunar module of the Apollo 11 mission, touched down in the south-west corner of the Sea of Tranquillity, an 80-kilometre expanse of dark volcanic rock towards the centre of the bright side of the Moon. The fact that one of the two men who climbed out onto the lunar surface was a physicist made a huge impact on me. When I later learned he had helped design the navigational equipment that helped get them there, I was hooked. Buzz Aldrin has to be one of the coolest PhDs on the planet – and, by the way, anyone who thinks the Moon landings weren't in earnest just needs to see the YouTube clip where Buzz punches out conspiracy theorist Bart Sibrel and I think they'll change their mind.

Just as the Apollo missions took us beyond our planet and out into the solar system, the Large Hadron Collider is going to take us somewhere we've never been before: the world of the small. We don't know what we'll find: Higgs particles, supersymmetric particles, additional spatial dimensions, maybe even little green men. But one thing is for certain: we are setting foot in new territory. The LHC is one titchy-tiny step for a man, a giant hormone-induced leap for mankind. I don't want to sound too sappy, but hopefully, like the Moon landings, it too will inspire a whole new generation of young people that maybe science is worth a shot.

Through the all-purpose detectors of ATLAS and CMS, the LHC is going to reveal the secrets of the quantum scale; through ALICE and LHCb, we are going to get a glimpse of a baby Universe, just a split second after the Big Bang. How that baby Universe evolved from the quark soup of the ALICE experiment into the fully-fledged, gargantuan thing it is today is the story of the next chapter. Buckle up, my friends, for we are going to glimpse the fascinating world of supernovae, the stars and the planets – and ask the question: if the Universe started with a bang, how is it going to end?

CHAPTER 3

WE ARE SLOWLY FALLING INTO AN ENORMOUS BLACK HOLE

On the last Wednesday in every month, as dusk falls, a rag-tag bunch gathers in London's Regent's Park. Their meeting point is the Hub, a hi-tech sports pavilion that could easily be mistaken for a crash-landed flying saucer. If you didn't know they were the Baker Street Irregular Astronomers, a loose collective of enthusiastic urban stargazers, you might think from the sea of bobble hats and earnest chatter that they were a support group for victims of knitwear. Some scour the sky with binoculars; others tinker with telescopes the size of water cannon; the vast majority, however, simply gaze up at the night sky without so much as a pair of spectacles. All of them, novices and experts alike, are there for one purpose alone: to commune with the wonders of the night sky.

You have to admire their tenacity. London isn't the ideal place to gaze at the firmament; for a start, there's too much stray light from the streets and buildings to make the stars stand out, which is why the Irregulars choose to meet in the middle of an unlit park. A bright moon can be a problem for the same reason, which is why their gatherings are scheduled well away from a full moon; that said, no amount of careful planning can account for the English weather. Tonight, for example, scanning the heavens feels less like tracking the gods and more like staring into a neon-orange bowl of lukewarm miso soup. All of which goes to show you why the world's finest telescopes tend to be located in Hawaii, Chile and the Canary Islands, where not only is there some very nice weather but plenty of mountains are on hand to raise the equipment up as high as possible and minimise blurring of starlight by the atmosphere.[1]

A star party, as we call it in the trade, is a great way to acquaint yourself with the heavens; not only are there experts on hand to point you in the right direction, but there's plenty of camaraderie to offset the stargazer's inevitable marrow-jangling existential angst. Because, as we shall see, not only is the night sky a window on some of the most breathtaking natural beauty you will ever witness, it also holds the key to what is arguably the greatest discovery mankind has ever made: the birth of the Universe. And if

---

1 The famous Hubble Space Telescope goes one better by hurdling the atmosphere altogether, and has been producing stunning images of the furthest reaches of the Universe for over two decades. If you've never visited before, you have a huge treat in store when you check out www.hubblesite.org . . .

that's not worth wearing a bobble hat in public for, then I don't know what is.

A BRIEF TOUR OF THE NIGHT SKY

Here's the best first step you can take in home astronomy. The next time it's a cloudless day, go and stand outside facing towards the Sun at the stroke of midday. If you are in the northern hemisphere, you will be facing due south. If you are ever lucky enough to own a house with a garden, this is the direction you want that garden to face, as it will then be a veritable sun-trap. Unless, of course, as happened to me once, someone buys the brownfield site at the far end of it and builds an enormous football stadium.[2]

Hold both your arms out straight, so that your left arm is now pointing east, where the Sun rose, and your right arm is pointing west, where the Sun will set. You now have a fabulous grid reference with which to view the night sky. A good compass app on your iPhone would have done much the same job, of course, but I want you to feel connected to the Sun and stars in the way that your ancestors did. Now let's play a few astronomical mind games that will truly plug you into the solar system, then move on out from there to explore the galaxy and the wider reaches of the Universe.

2 Arsenal, in case you're interested.

HOME SWEET HOME

Let's pick up where we left off, facing south in daylight.[3] Now trace the Sun's path across the sky from east to west; this imaginary line shows you the plane of the solar system. Astronomers call it the *ecliptic*, because when the Moon hits this line it's game on for an eclipse. Everything in orbit around the Sun passes through the sky on roughly this arc. Only the Moon is usually bright enough to make an appearance in daylight, but when night falls our five neighbouring planets – Mercury, Venus, Mars, Jupiter and Saturn – can often be seen rising in the east and setting in the west, all following this same curved path.

The Moon and the planets don't produce any light of their own; they are just reflecting back the light of the Sun. At first glance they look a lot like stars, though if you were to keep watch over a period of hours you would see that they are moving relative to the starfield. The giveaway sign is their relative brightness, and the fact that, being so much closer, they are less affected by interstellar medium and so don't twinkle. The three that are further out from the Sun than us – Mars, Jupiter, and Saturn – tend to be the most visible at night. The other two, Mercury and Venus, are close to the Sun, so tend to be easiest to spot at dusk and dawn. Of the two, you are far more likely to spot Venus, the breathtaking 'Morning Star'; after the Moon, it's usually the brightest thing in the sky.

---

3 If you are in the southern hemisphere, of course, you will be facing due north, and estate agents will be bragging about north-facing gardens instead.

Mankind has always measured time by the heavens, so it's perhaps no coincidence that the seven days of the week are named after the Sun, Moon and five closest planets. Sunday and Monday (Moon-day) are obvious enough; the other planets are named after Roman gods and, if your Latin is as bad as mine, easier to cipher in French: *Mardi* is Mars-day, *Mercredi* is Mercury-day, *Jeudi* comes from Jupiter-day, and *Vendredi* from Venus-day. And Saturday, everyone's favourite respite, is quite fittingly a lazy way of saying Saturn-day. The Anglo-Saxon variants require a little more parsing: Tuesday comes from *Tyrsday*, with Tyr being the Mars-like Norse warrior god. Odin, the Norse god linked with Mercury, gives us *Odinsday*, now Wednesday. The Norse god Thor, a dead ringer for Jove of Jupiter fame, winds up as Thursday; and Freya, Norse fertility goddess and Venus lookalike, has her namesake in Friday. You see? If it weren't for the Vikings it would all make much more sense.

Uranus and Neptune are out there too of course, though being the two planets furthermost from the Sun, they are the hardest to spot, usually needing a pair of binoculars and a bit of expert guidance to pick them out. Locating Pluto, on the other hand, requires a powerful telescope and a professional astronomer, and in any case it has recently been demoted from planet status to that of a 'dwarf planet'. That may seem cruel, but as we shall see it was a necessary step, given the fact that all manner of other similarly sized rocks were turning up in roughly the same orbit and there quite simply weren't enough names of Roman gods to go round.

SATELLITES, METEORS AND COMETS

So you know the points of the compass, and the exact path that the Sun, Moon and planets trace through the sky, day and night. In astronomy terms, you have eaten your solar system greens and are now ready for pudding. Lucky you, because the night sky is jam-packed with all manner of delicious treats.

First, of course, there's the incredibly impressive man-made stuff. Since the debut of Sputnik in 1957, we have launched over three thousand satellites, and they can make for spectacular viewing. They orbit close to the Earth, so the best time to see them is well away from city lights, at dusk or dawn at the time of a new Moon when there is no moonlight to drown them out; after all, you are looking for light reflected off something the size of an articulated lorry a few hundred kilometres away. Once you get your eye in, you will be amazed by how many you can spot. You need to be quick, though; my personal favourite, the International Space Station, skips across the horizon in about five minutes.[4] Incredible, isn't it? A real space station, with real cosmonauts on it, drifting around the Earth so close that you can see the huge solar panels that keep their stereotypical vodka chilled.

Communications satellites like the one that brings you Sky TV are too far away to be visible with the naked eye, parked as they are in a stationary orbit some 36,000 kilometres from

4 There is an amazing website, www.heavens-above.com, where you can track any satellite you care to name and get a tip-off when it's likely to cross your own particular patch of sky.

Earth; navigational satellites like the ones that run your satnav are not much closer at some 20,000 kilometres; but all the interesting stuff like space stations, telescopes and meteorological satellites tend to be in what's called a Low Earth Orbit, within 2,000 kilometres and well within range of your unassisted peepers.

For me, however, by far the most exciting home-grown objects you can ever see in the night sky are the solar system's free fireworks: meteors and comets. Meteors, as you may know, are basically bits of space rock burning up as they strike the Earth's upper atmosphere, seen as a bright flash in the night sky that lasts for a few seconds before fading away. Comets, on the other hand, can be visible for months at a time, and are huge dirty snowballs up to several kilometres in diameter, which heat up as they approach the Sun to form a huge fireball, trailing colossal plumes of gas and dust. Not only are both of these space visitors ridiculously impressive to watch but, as we shall see, they provide vital clues as to the origins of the solar system and may even hold the key to life on Earth.

The theory goes like this. The Sun was born some 5 billion years ago, along with multiple other stars, as a gravitational clump in a 'stellar nursery' of hydrogen gas and old stardust. As the Sun became more dense, its gravity got stronger, and it dragged towards it all the dust, gas and ice that it could get its tiny stellar hands on. There must have been some slight rotation in this swirling mass of star-and-planets-to-be, because as it condensed under gravity, it speeded up to form a disc, much in the same way as a skater spins faster when she draws in her arms. This is the reason that our Sun is spinning, and why all the planets are orbiting it in the same direction. As the Sun's gravity hoovered up the last bits of neighbouring hydrogen, the denser regions of the disc slowly began to clump together under gravity to form planets. Finally, the Sun ignited and began to burn, producing a solar wind that blasted away all the stray gas from the solar system, forcing the Sun away from its sibling stars.

The planets in the disc formed into three main sequences. First, there was the inner belt of small rocky planets: Mercury, Venus, Earth and Mars; next came the two gas giants, Jupiter and Saturn; and making up the rear were the so-called ice giants, Uranus and Neptune. Not all the material in the disc formed planets, however. Between Mars and Jupiter, there's a sort of junkyard of failed rocky planets and general rubble called the Main Asteroid Belt, and this is thought to be the source of most of the stray space rock that ends up as meteors. As you probably know, asteroids are going to be big business over the next couple of decades. In 2010, President Obama decided to decommission the Space Shuttle and set NASA a new challenge: to land a man on an asteroid by 2025. At first glance this might seem a little bit lame in comparison with a Moon landing, but the more we've learned about asteroids, the more important they seem to be.

For a start, there's the very real danger that a wonky one might put us all out of business. Most meteors are made of stuff the size of a pebble, but once in a while a truly heavyweight asteroid finds its way into a collision course with the Earth, and enough of it is left over after burn-up to cause some really serious damage. Try this for size: we now believe the mass extinction of the dinosaurs at the end of the Cretaceous period, some 65 million years ago, was caused by an asteroid over 10 kilometres across, leaving behind it a crater 180 kilometres wide in Chicxulub, Mexico. Try keeping your teacup in its saucer the next time one of those bad boys hits town.

So President Obama has a point. If we can learn to manoeuvre ourselves onto one of the many large rocks that occasionally buzz the Earth, we have a fighting chance of diverting a stray one if it ever threatens to pay us a visit. And, equally importantly, asteroids seem to be the missing link in one of the most intriguing puzzles of Creation: if the Earth formed from a collapsing plate of red-hot dust, where did its water come from?

WATERY ASTEROIDS

One look at the pock-marked face of the Moon tells you it's no stranger to the odd asteroid collision, and the Apollo missions surprised everyone by showing that all its craters were made in a relatively narrow window some 4 billion years ago, about 600 million years after the solar system first formed. It's now thought that the Moon was pummelled by a mass migration of asteroids in a cataclysm called the Late Heavy Bombardment. And if the Moon was bombarded, the thinking goes, then so were the terrestrial planets, the Earth included. There's very little geological activity on the Moon, so the craters have stayed put, whereas on the Earth and other nearby planets like Mars and Venus they've been smoothed out due to volcanic eruptions and plate movement.

Several asteroids have been discovered recently that are carrying enormous quantities of water; some of them have enough to fill the Earth's oceans several times over. An intense bombardment of a cooling Earth by soggy asteroids goes a long way to explaining not only where our water came from, but also why, even though the Earth was formed about 4.6 billion years ago, we can't find any rocks that old.[5] If our theories are correct, a Late Heavy Bombardment by watery asteroids roughly 4 billion years ago melted all the Earth's newly formed surface rock, creating towering clouds of steam that were captured by the Earth's gravity, eventually condensing as the oceans. In other words, the water in your glass may well have come from beyond Mars.

WHAT KIND OF COWBOYS BUILT THIS SOLAR SYSTEM?

The rocky detritus left over from the formation of the Earth-like planets is nothing, however, compared to the vast swathes of icy rubble left over after the formation of the ice giants Uranus and Neptune. Just as the rocky junk outside Mars is the main source of meteors, the icy junk outside Neptune is the source of comets. Some of it sits in a roughly doughnut-shaped region called the Kuiper Belt; this is home for Pluto and a few other wannabe planets. Most of our shorter-period comets –

---

5 You may be wondering how we managed to date the Earth, given that none of the first rocks have survived. One of the ways is to look at the composition of meteorites, which formed at the same time as the Earth but are too small to be geologically active.

the ones that pay a visit every few years – are thought to originate from here. The longer-period ones are believed to originate even further out, in a mysterious bubble of icy debris called the Oort Cloud that surrounds the entire solar system nearly a whole light year out from the Sun.

Yes, that's right: a whole light year. That's a thousand times further out than the far edge of the decidedly out-in-the-sticks Kuiper Belt, and a quarter of the way towards Proxima Centauri, our nearest star. Indeed, it may be that many of the comet nuclei in the Oort Cloud were stolen from the planetary discs of other infant stars back in the days of the stellar nursery. In other words, the comets we occasionally glimpse in the night sky could well be visitors from alien worlds. They may even have been responsible for bringing life to Earth in the first place. After all, if comets can move material around from one baby star to another, the creation of microbial life might be a rare event that spreads among infant solar systems in the stellar nursery, in much the same way that chickenpox soon gets around at a kindergarten. Thanks to asteroids and comets, the Earth is far from isolated; it is directly connected with the very furthermost reaches of the solar system and possibly even the stars beyond.

THE ZODIAC

So now that we've got a feel for what's going on in our own backyard, let's take another look up into the sky and check out our nearest neighbours, the stars. They are all so far away, of course, that our depth perception is flummoxed and the starfield as a whole appears to hang over us like a dome; in reality stars differ immensely both in their intrinsic power and in their distance from the Earth. Alpha Centauri, neighbour of Proxima, is the nearest star that's visible with the naked eye, and its light takes about four years to reach us.[6] Deneb is one of the furthest visible without any form of optical jiggery-pokery and its light takes roughly 1,600 years to make the trip. Which means, when you think about it, that Alpha Centauri appears as it did before the last Olympics, whereas we are seeing Deneb round about the time of the first Sack of Rome.

The brightest stars have traditionally been grouped into con-stellations that bear − if you ask me − extremely flimsy resemblances to a motley collection of animals, gods and house-hold objects. The Greeks named the majority of the northern ones; as far the southern menagerie goes we have only ourselves to blame. A great number of the animal signs lie along the ecliptic, which was a big deal for the Babylonians and gave birth to the Zodiac, as in *zoon*, Greek for animal. If you are a pas-sionate believer in astrology, however, don't be disappointed if the Sun appears to be in the 'wrong sign' at noon on your birthday. As we shall see later in Chapter 7, 'The End of the World is Nigh', the Earth has a slow wobble on its axis, making one full cycle roughly every 26,000 years. This means that, from the perspective of the Earth, the constellations have shifted back by about one star sign over the last 2,000 years. In other

6 I should point out that Alpha Centauri is only visible from the southern hemisphere, in the constellation of Centaurus, so don't strain your eyes looking for it from The Hague.

words, if everyone has been telling you you're a leader-of-men Leo, you're really a shy retiring Cancerian.[7]

The International Astronomical Union – the same killjoys who demoted Pluto to a dwarf planet – define eighty-eight constellations, which, although they look nothing like the things they are named after, do at least form some sort of grid reference for the sky. Before we move beyond our neighbouring stars and explore the twinkling citadel of our galaxy, let's just pick out one or two stellar treats that will really perk up those long moonless country evenings when you're forced to walk home alone from the village barn dance.

LIKE A DIAMOND IN THE SKY

So that we know what we're dealing with, let's just briefly recap what we know about stars from Chapter 1. When a cloud of interstellar gas and dust and ice suffers some sort of shock wave – caused, for example, by a nearby star exploding in a supernova – some bits become more dense than others and start to collapse under their own gravity. The chances are that the shock wave will also cause a bit of rotation in the clumps, and as they contract they will start to spin faster, forming a central ball of hot collapsing gas surrounded

---

7 Oh, and when you do look up into the night sky to find the constellations that line the ecliptic, you'll also notice there's actually a thirteenth constellation, Ophiuchus, the Serpent Bearer, in between Scorpius and Sagittarius. In case you're wondering, Ophiuchuses are 'passionate believers in astrology who are able to explain away any hard evidence that in any way contradicts their barmy world-view'.

by a disc of material that will eventually form planets. Soon the central ball of gas gets sufficiently dense and hot for nuclear fusion to switch on, turning hydrogen into helium and releasing enormous amounts of energy. A baby star has formed.

When stuff gets really hot it starts to give off visible light. Most stuff radiates red light by the time it reaches a temperature of about 1,000 K (roughly 700°C), which is where the expression 'red hot' comes from. And I really do mean anything: it doesn't matter a jot whether it's a lump of coal or a banana. Keep heating that thing till its temperature reaches 6,000 K and it will start to emit white light; raise its temperature even further to about 10,000 K and the light it gives off will be blue. Not that this is news, really: after all, when we light a fire in the grate, we all instinctively understand that a white-hot ember is hotter than one that glows dull red, and that a blue flame is hotter than a white one.

THREE OF A KIND

So what has this got to do with stars? Well, stars are, at their most basic, enormous nuclear fusion reactors that are so hot that they are giving off visible light. Blue stars are the biggest, hottest and brightest, and tend to 'live fast, die young'. They can be anything from about ten times the mass of the Sun to a hundred, and they stick around for a relatively short time – millions of years or so – before they exhaust their supply of nuclear fuel and explode as supernovae. All the atoms you are made of came from truly spectacular landmark stars like this. There have been countless generations of them since the dawn of Creation, in a repeating cycle: shock wave from supernova creates stars, the big stars among them explode as supernovae, shock wave from supernova creates more stars.

White stars, like our Sun, are the most common births in these nurseries and have middle-ranging sizes and temperatures.[8] A star like this lives for a few billion years before bloating as it runs out of fuel. Once nuclear fusion switches off altogether, the outer layers of gas and dust drift away, forming a cloud called a planetary nebula, while the core shrinks down to something the size of the Earth. Astronomers call these dense Earth-sized objects 'white dwarfs'. The core usually gets as far as making carbon, so a star like the Sun really does end up as a diamond in the sky, albeit one that's white hot and the size of the Earth.

At the other end of the spectrum, red stars are the smallest, coolest and faintest, and they live for trillions of years. The Universe itself is thought to be only 14 billion years old, so, if you think about it, every small red star ever made is still alive, even the ones created at the very dawn of the Universe. The naming of stars pre-dates any form of political correctness; therefore stars bigger than the Sun often go by the name of

---

8 I know what you're thinking: the Sun isn't white, it's yellow. In actual fact, it appears yellow only because of our atmosphere. Air molecules tend to scatter light, and the bluer the light the more it gets scattered. All the blue light from the Sun therefore ends up bouncing around in the sky, making the Sun appear yellow and the sky appear blue. If you were to look at the Sun from space, you'd see its true colour: a dazzling, brilliant white.

'giants' and 'supergiants', while stars the same size as the Sun and smaller are rather indelicately called 'dwarfs'. Small red stars, or 'red dwarfs' as they are known, just eke out their supply of nuclear fuel and, as far as we know, never get much further than converting hydrogen to helium. They seem very uninteresting indeed, until you consider that any life form that manages to survive on a planet orbiting a red dwarf would have it made, as their home star would, to all intents and purposes, never burn out. More on that later in Chapter 8 when we come to discuss extraterrestrial life . . .

BOLD GIANTS AND SHY DWARFS

So stars are made in a stellar nursery of collapsing hydrogen gas, and come in three sizes: large (blue), medium (white, like the Sun), and small (red). Now let's think about what we might expect to see when we look up into the night sky.

First of all, we'd expect to see lots of small red stars, because they have such ridiculously long lifespans, as well as lots of medium-sized white stars like the Sun, because they get made the most often. We'd expect to see very few big blue stars, because they are rarer and burn themselves out so quickly. One glance in the night sky, however, and you see hardly any red stars at all; nearly all of them are either white or blue. So what's going on?

The answer, of course, is that the red stars are out there in roughly the proportions we'd expect, but we just can't see any of them with the naked eye. Proxima Centauri is a red dwarf, but even though it's our nearest neighbour it's too faint to see without a powerful optical telescope. And it's strictly southern hemisphere, of course, so you won't see it at all if you live north of the equator. In a similar way, even though blue stars are rare, they are so bright that they tend to be dispropor-tionately easy to see with the naked eye. Imagine that: as many as 10 per cent of the stars that you see in the night sky are enormous and as rare as hens' teeth.

HERE COMES THE SUN

We've already mentioned our nearest Sun-sized star, Alpha Centauri; those in the southern hemisphere will know it as the brighter of the two Pointer stars that lead the eye to the Southern Cross. For those in the northern hemisphere, the closest Sun-like star is Sirius, also known as the Dog Star. It's about twice the mass of the Sun and twice the size, and a mere 8.6 light years away. It's usually the brightest star in the north-ern night sky; just look for Orion the Hunter, and Sirius is the bright star snapping at his heel.

Now let's turn to an elderly Sun-type star that's entered it's bloating phase, also known as a 'red giant'. It's called Arcturus; to find it just look for the Plough, the saucepan-shaped group of seven bright stars, and extend the arc of the 'handle' away from the 'bowl'; it's the next bright star you come to. At thirty-six light years away, it's about five times further off than

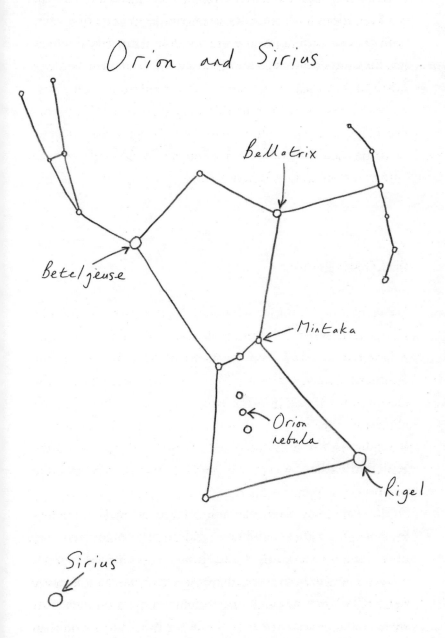

Orion and Sirius

Bellatrix

Betelgeuse

Mintaka

Orion nebula

Rigel

Sirius

Sirius, and though it has pretty much exactly the same mass as the Sun, it's swollen to about twenty-five times its diameter. One day our own fair Sun will end up looking very much as this Elvis-on-the-toilet star does now.

The Plough and the North Star

North Star
aka
Polaris

The Plough

Arcturus

SOMETHING BLUE

So those are some classic examples of Sun-like stars. Now let's look at some rather chunkier – and hotter – blue stars. Not only is the constellation of Orion one of the easiest to find, but it's visible in both hemispheres and every star in it is enormous. The first thing to look for, whichever hemisphere you are in, is the three horizontal stars of the belt; hanging directly below them – if you are north of the equator; above it if you are south – are the three stars of the sword. They could have called it something else, but they went for sword. Unbelievably enough, the middle of those three sword stars is in fact the Orion nebula, your own personal stellar nursery. It's about 1,300 light years away and twenty-four light years across, and, within its clouds of hydrogen and dust, baby stars are forming.

Next let's look at a really large blue star in short trousers. Go back to Orion's belt; the furthest to the right of the three stars is Mintaka, and you'll see it burns with a fierce blue light. Although it's the faintest of the three belt stars, it's about twenty times the mass of the Sun and about 90,000 times as bright. The only other spring chicken among Orion's clutch of enormous blue stars is Bellatrix, the hunter's left shoulder; the others are all blue supergiants, meaning they have stopped burning hydrogen and have swollen to nearly one hundred times the diameter of the Sun as they make heavier elements. Rigel, Orion's right foot, is a classic example.

Now let's rubberneck a really big star near the very end of its life. Find the three stars of the belt again, and then look up to Orion's right shoulder. That ruby-red star is the supergiant Betelgeuse – pronounced 'beetle juice' – and it is expected to blow up any time in the next million years or so. The blue supergiants in Orion like Rigel are big, but Betelgeuse, at roughly ten times the size of the Sun, is a whole other story. It has pretty much exhausted its supply of nuclear fuel, and is working on converting as much of its core as it can to iron, the heaviest element that gets made in any star. Once fusion switches off completely – as it could when you are watching it tonight – the core of the star will collapse under gravity, releasing enormous amounts of energy, which then heat the outer layers of the star until they explode in a supernova.[9] At about 600 light years away, it's comfortably out of range; even so, when it goes it will be so bright that it will easily be visible in daylight. Its shock waves will spread out into the surrounding interstellar gas, causing new clumps of matter to form, which will then collapse under their own gravity to form a whole new generation of stars.

Betelgeuse is so big that, once it has exploded, whatever gets left behind will probably be too big to form one of those Earth-sized lumps of white-hot carbon also known as a white dwarf. Instead it will suffer one of two fates. If what gets left over is larger than about 1.4 solar masses, it will squish down under its own gravity to form a ball of stupendously dense matter known as a neutron star. A white dwarf is dense, but a neutron star is about a billion times denser; so much so that just a pinhead's worth would weigh about 1,000 tonnes. On the other hand, if Betelgeuse's supernova leaves behind a core of more than roughly three solar masses, it will crush down to form something so dense that it makes a neutron star look like candy floss: a black hole.

---

9 Since only atoms as big as iron get made in the cores of stars, you may be wondering how the rest of the ninety-two elements we find on Earth get their start. The answer is that they are made in the incredibly high pressures and temperatures of supernovae. In other words, the iron in your steel fork was probably forged by nuclear fusion in a huge blue star; the lead in the flashing on your roof was seeded in a colossal stellar explosion.

THE DREADFUL FATE OF SUPERSTARS

Back in Chapter 2 we talked about how collisions between particles in the LHC might create black holes, and how they wouldn't – by definition – be directly observable, but would probably show up as a splurge of light as they almost instantaneously evaporated. Needless to say, black holes created by squishing down massive stars aren't nearly so benign. At a respectful distance they would behave just like any other large object: if you angled your spaceship just right, for example, you could orbit safely around a black hole just as you could around any large star. Get too close, however, and you would find yourself in the grip of a gravitational field so strong that not even light could escape.

How close could you safely go? Well, the event horizon, as it's called, is the point of no return; for a black hole with a mass ten times that of the Sun, we're talking roughly 30 kilometres away. In practice, you wouldn't be able to get anything like as close as that, because gravity would be so strong that it would tear you and your spaceship apart at something like ten times that distance. To be on the safe side, you'd need to set an orbit of just over 3 million kilometres, which is about a fiftieth of the distance of the Earth from the Sun. That would give you a ride no more stressful than Space Mountain at Disneyland, which was quite enough for me.

X-RAY SPECS

So if we can't see black holes, how can we be sure they are out there? After all, if even light can't escape them, they aren't exactly going to show up with a telescope, are they?

The answer, interestingly, is that they don't show up with telescopes that use visible light, but they do show up if you use telescopes that look at other wavelengths. There's more to light, of course, than just the bit we can see: X-rays, ultraviolet, infrared, microwaves and radio waves are all essentially the same stuff; we just can't detect them with our eyes. With the right kit, we can make images using these other wavelengths in much the same way that our eyes make images using visible light. And radio waves and X-rays turn out to be the tell-tale signs that a distant region of the galaxy is harbouring a black hole.

Black holes, it turns out, are very messy eaters. Most stars, as you know, are born in clusters in stellar nurseries, and they frequently get tangled up together by their mutual gravitation. Our Sun is a bit of a loner;[10] roughly a third of the stars we see around us have one or more partners of varying sizes. Polaris, for example, is actually orbited by no less than four other stars. Often a stellar black hole will start to feed on its siblings, with matter streaming from the outer layers of the victim star and

---

10 At least, as far as we know. There's actually a theory that somewhere, off beyond the graveyard of icy rubble in the Oort Cloud, is another small faint star, twinned gravitationally with the Sun, which has yet to be discovered. This hypothetical companion has even been given a name, Nemesis, and the idea is that it's responsible for periodically kicking lumps of ice out of the Oort Cloud and so creating long-period comets.

radiating enormous amounts of energy as it is sucked in. This energy comes off in the form of X-rays and radio waves, and, with some clever detective work, it can point the finger at something that just has to be a black hole.

THE DARK HEART OF THE SWAN

Remember Deneb, which at 1,600 light years away is one of the most distant stars visible to the naked eye? Well, Deneb is the 'tail' star in the northern constellation of Cygnus, or the Swan. In its midriff is an object called Cygnus X-1, which is believed to be a star-sized black hole feeding on a thirty-solar-mass blue supergiant. Of course, we can't see the black hole directly; instead we can see material streaming from the companion star into a small invisible object, releasing huge quantities of X-rays in the process. By measuring how long it takes for the feeder star to complete each orbit, we can tell that the invisible object is about ten times the mass of the Sun: in other words, much too heavy to be a neutron star and in all probability a 100 per cent genuine black hole.

Cygnus is one of the easiest constellations to find in the summer months, and it dominates the northern skies in the early evening in much the same way that Orion does in the winter. Just look for the famous 'summer triangle' of bright stars, Deneb, Vega and Altair; they are the three brightest stars in the south-east and form a downward-pointing arrow. The top left corner of the triangle is Deneb; now look more closely

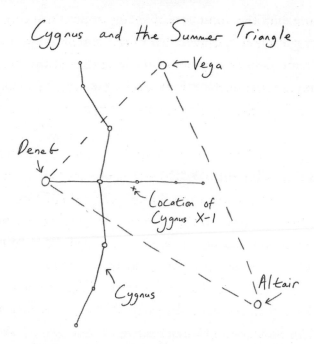

Cygnus and the Summer Triangle

and you will see it become the tail of a swan in flight, with its wings outstretched and its neck pointing directly between Vega, at the top right, and Altair, at the bottom. In the middle of that small group of bright stars is one of the darkest, densest objects in Creation. Thankfully, at some 6,000 light years away, it's even further off than faraway Deneb. Even more thankfully, a different solar system is providing lunch.

THE RIVER OF LIGHT

With all this talk of stellar black holes way beyond Deneb, we are now looking out into the wider reaches of the galaxy and nothing but the most sensitive telescopes, working with every available frequency of light from X-rays to radio waves, is going to help us. But before we leave the galaxy altogether and explore the furthest limits of the Universe, let's take one final look in the night sky with nothing but our unaided eyes.

We're on first-name terms now, us and the firmament. We've watched the five closest planets of our own solar system wander across the ecliptic. We've singled out Alpha Centauri, the nearest visible star. Orion, Cygnus and the Plough are our wingmen. We know that, by and large, the stars of the canopy are big game: huge pyres of carbon, hydrogen and every element in between, each one of them many times more massive and bright than our own humdrum Sun. Many of these giants and supergiants, we now know, will end their days as stunning supernovae, showering the nearby Universe with every element in the periodic table, and their embers will crush down under their own impossibly enormous weight to form white dwarfs, neutron stars or even black holes. But there's one unimaginably enormous object in the night sky that we haven't got to grips with. The Milky Way.

There are some 4,000 stars in the starfield that are technically bright enough to be seen without any form of optical leg-up. Stare up on a clear moonless night, however, and the impression is of a countless multitude, thanks mainly to a twisting river of ghostly light that spans the entire heavens. The light is from stars in their billions: over 100 billion, in fact. It is an extraordinary truth, but our galaxy is actually a thin disc, with a bulge at the centre,[11] with the majority of its stars contained in a region roughly 100,000 light years across and only 1,000 light years thick. In other words, when we look at the Milky Way, we see a strip of light in the sky, because we are looking directly along the plane of the disc.

The appearance of the Milky Way in the night sky tells us an awful lot. For a kick-off, at its zenith the Milky Way runs right across the middle of the sky. That means, in terms of depth, we must be about halfway down into the disc, with roughly as much Milky Way above us as beneath. And this is borne out by measurement: so far as we can tell, the Sun is currently about 100 light years above the plane of the galaxy and heading for the top edge.

Second, the Milky Way doesn't lie on the ecliptic. If it did, that would mean the solar system lay on the same plane as the galaxy. Look to where the Milky Way and the ecliptic meet, and you'll see they make an angle of about 60°. That means – you guessed it – the solar system isn't 'flat' in the plane of the disc, but instead sits at a jaunty 60° angle.

Lastly, of course, the Milky Way is thinner in some parts than in others. If it was the same width all across the sky, we'd know we were slap bang in the middle of the galaxy. As it is, we know we must be some way out from the centre; in fact our best estimate is that we are about halfway between the hub and the edge. Generally speaking, the northern hemisphere looks out towards the edge of the disc, and so the Milky Way

11 Actually, if you could look down on the Milky Way galaxy from above, you'd see it was a so-called 'barred spiral'. In other words, the stars aren't spread out uniformly from the centre. Instead there's a bar in the middle, trailing spiral arms, a bit like a Catherine wheel. Our own solar system is located in what's called the Orion Arm, with Orion on one side and Cygnus on the other.

seems thinner and fainter. The southern hemisphere, on the other hand, faces the centre, so the Milky Way appears wider and brighter. The exact centre of the disc happens to lie in the constellation of Sagittarius; in other words, the stars of Sagittarius hang like a net curtain, through which you can see the clot of stars at the hub of the galaxy.[12]

And here's the kicker. You can feel the gravitational pull of the Earth, keeping you permanently stuck to its surface. You are also aware of the pull of the Sun, forcing the Earth – and you with it – into a circular orbit so large that it takes a year to complete. But you may not have realised that you, your cat and the entire solar system are trapped in an orbit around an even bigger beast, taking some 250 million years to make a round trip. That beast lurks right at the heart of the galaxy, in that cloud of ethereal stars behind Sagittarius. It has its grip on you right now, whether you know it or not. It is a stupendously large black hole, and it has a mass 4 *million* times that of the Sun. And at the end of this chapter, we're going to fall right into the middle of it.

A SELF-MADE MAN

If you think that the greatest scientists have all been invertebrate geeks, then you need to know about Edwin Hubble.

---

12 Sagittarius, of course, is mainly a southern constellation. If you're in the northern hemisphere, don't feel hard done by: you can console yourself with a great view of the outer edge of the disc. Find Orion, then head north towards the pole star. Very quickly you'll hit the Milky Way. You'll see that it's very thin and faint, because you are now looking straight out at the edge of the galaxy.

Not only was he arguably the most important astronomer since Galileo, but he was also a chisel-jawed athlete who would have creamed Ernest Hemingway in a fist fight. And what is more, he has the rare distinction among scientists of having changed the way we think about the entire cosmos, not once, but twice.

If you had met the flamboyant Rhodes Scholar during his three years at Oxford from 1910 to 1913, you might have assumed from his upper-class English accent, cape and cane that this young law student from Queen's College was an aristocrat of the fruitiest European stock. Everything about Hubble, from his impressive height and athletic ability to his quiet self-confidence, screamed privilege and breeding. In reality, he was from a long line of small-time Missouri farmers and as apple-pie-American as they come. Prone to tall tales of drawing-room duels and heavyweight boxing matches, and virtually self-taught as an experimental astronomer, Ed Hubble was in all respects a self-made man.

Once he had got the pleasures of the athletics track and the party circuit out of his system, the young Hubble made phenomenal progress. He became fascinated by a particular kind of bright cloud called a 'nebula'. It was thought at the time that the Milky Way galaxy was all there was to the cosmos. Hubble, working long cold nights with the new 100-inch Hooker Telescope on top of Mount Wilson in southern California, proved otherwise. Some nebulae, he showed, weren't simply bright blobs of nearby gas;[13] they were entire galaxies, just like our own Milky Way, and millions of light years away into the bargain. Ed Hubble had discovered the Universe.

Many people would have left it there, but Hubble, as you are probably beginning to guess, was not most people. He found galaxy after galaxy, and examined the type of light they gave off. Now it just so happens that every atom and molecule has a characteristic 'bar code' in terms of the light that it emits. A sodium street light, for example, produces a particular shade of yellow. Hydrogen, on the other hand, which is much more plentiful in galaxies, gives off a characteristic frequency of ultraviolet. Hubble noticed that these 'bar codes', or *spectra*, tended to be 'red-shifted'. In other words, they had shifted towards the red, or lower-frequency, end of the spectrum.

If the light from something is red-shifted, it generally means that the thing is moving away. As you'll know if you have

13 Hubble did this using a certain kind of star called a 'Cepheid variable', after the first recognised star of its kind, Delta Cephei. A Cepheid variable has finished burning hydrogen and has started working on its helium, and it has swollen from 'dwarf' size to 'giant' in the process, burning ever so slightly brighter and darker with a regular cycle as it undergoes a kind of 'tug of war' between radiation and gravitation. These stars are the distance markers of the Universe. The time it takes for a Cepheid variable to go through each cycle tells you how intrinsically bright it is, and from its apparent brightness in the sky you can therefore work out how far away it is. Neat, eh?

Even neater is the fact that our own North Star, Polaris, is a Cepheid variable. Many people expect it to be the brightest star in the northern sky, but it's actually quite faint. To find it, look for the Plough, focus on the two stars that make up the far side of the 'bowl', trace a line from the star at the bottom of the 'bowl' to the one at the top, imagine extending that line about five times and you'll come to Polaris. It will be the same height above the horizon that you are above the equator, so if you're pointing up at it from Los Angeles, your arm will make an angle of 34° with the horizontal; if you're at the North Pole it'll be right over the top of your head. Polaris looks white to my eye; it's about six times the mass of the Sun and about 430 light years away. Watch the sky throughout the night and the entire heavens will appear to rotate about it like a giant stellar tumble-dryer.

stood beside a train track as a passing train sounds its horn, the pitch of the horn is higher as the train approaches you, and lower as it departs. That's due to something called the Doppler effect, which says that if a wave source is moving towards you, its waves will get bunched up and therefore have a higher frequency; if it's moving away, the waves will get more stretched out and their frequency will be lower.

Light is also a wave, so the fact that the light given off by hydrogen in distant galaxies was lower frequency than it should be – in the visible, say, rather than the ultraviolet – meant that those galaxies were all moving away from one another. Not only that, but Hubble found that the further off the galaxy, the more red-shifted its light and therefore the faster it was retreating. The Universe, Hubble realised, is not static. Not only are all its galaxies moving away from one another, but the further off they are, the faster they are going. Run the film backwards, of course, and the whole kit and caboodle must have started from a single point. The Universe must have had a beginning. Hubble had discovered the Big Bang.

TWIN PARADOXES

Let's go back to the night sky, because it's time to make one final naked-eye observation that ties in beautifully with Hubble's expanding Universe. Sometimes it's easy to miss the obvious, and it doesn't get more obvious than this: the night sky is black.

Let's think about what that means for a moment. If the

Universe is infinite, as astronomers traditionally believed, then there are an infinite number of stars within it. That means, in every direction you look, there's a star. Light travels in straight lines, so surely the night sky should be a dazzling white, not pitch black?

This pickle is called Oblers' Paradox, after the German astronomer who popularised it in the mid-nineteenth century, but it had been niggling astronomers for centuries before that. Hubble's discovery implies that the Universe had a beginning, and therefore supplies a neat solution. After all, if the Universe has a finite age, then the light from the most distant stars will not have had time to reach us, and the night sky will be black, not white. Put another way, the fact that you can look up into the night sky without having to wear shades is supporting evidence for the Big Bang.

Of course, the expression 'Big Bang' is unhelpful in one sense, because it immediately conjures up an image of galaxies exploding out from a central point. In which case, what are they exploding out into? Luckily, one bright spark was keeping a close eye on Hubble and his results, and managed to provide a few answers ...

ALBERT EINSTEIN

Education, like a lot of institutions in life, tends to be a 'one size fits all' system, which suits most of us just fine but doesn't seem to work quite so well for people at the extremes. Albert Einstein was famously one of those people. The myth paints him as a late developer and high-school drop-out, lobbing intellectual grenades at the scientific establishment; the more interesting reality is that his education was quite simply unremarkable. At both primary and secondary school he showed ability in science and mathematics, and trained as a teacher at Zurich Polytechnic as a way of funding himself as a graduate student. Failing to find a teaching post, he quite sensibly took a day job as a junior clerk in the Swiss Patent Office, enabling him to study in his spare time for a PhD in physics at the University of Zurich.

So far, so humdrum. But immediately after gaining his PhD in 1905, Albert had what can only be described as a very good year. Seemingly out of nowhere, he published four ground-breaking papers, any one of which might be considered worthy of a Nobel Prize. Two of them laid the foundations of quantum mechanics, the first by proving the existence of atoms and the second by showing that light consisted of particles or 'photons'. And the other two established what is possibly the best-known physical theory of the modern age: General Relativity.

As we glimpsed in the last chapter, there is a bit of an impasse in contemporary physics. Put crudely, we have a very good theory of big things and an equally impressive theory of small things, but they are both so different that it is hard to see how they might be made consistent with one another. Where small things are concerned we have the Standard Model, which manages to combine the weak, the strong and electromagnetic forces, but sweeps gravity under the carpet. And for big things we have General Relativity.

And by big things I really do mean big, because – incredible as it sounds – General Relativity actually provides an equation that describes the Universe. It ignores the strong and the weak forces, and instead concentrates on the relationship between gravity and electromagnetism. Mathematically it's a bit of a nightmare, with solutions of the equations being notoriously hard to find, but several gifted individuals have managed to show that it predicts an expanding Universe full of black holes, just like the one our telescopes tell us is out there.

EVERYTHING IS RELATIVE

You may need to be a number cruncher of the highest rank to solve the set of equations that comprises General Relativity, but anyone is capable of understanding the simple ideas on which it's based. Albert Einstein, as you shall see, had an extraordinary gift whereby he would take a simple insight and follow it to its logical conclusion, no matter how left-field that conclusion appeared to be. In the case of General Relativity, his simple questions led him to abandon our common-sense understanding of time and space, and to reject altogether the idea that gravity is a force.

The simple insight that he started from was this. Imagine that you are standing in a stationary elevator, with no windows, parked on the surface of the Earth. If you were to do a few experiments, like timing the swing of a pendulum or bouncing a ball, any measurements you made would all lead you to the conclusion that you were in the Earth's gravitational field. Now imagine your elevator is out in deep space and the gravitational field around you is zero. If the elevator were accelerating at just the right rate, your experiments would give you exactly the same results. You would feel a force pulling down on you, just as you did when you were parked on the Earth's surface. This time the force is due to the elevator's acceleration rather than the gravitational pull of the Earth, but you wouldn't be able to tell the difference. Likewise, if you conducted a few experiments, you would notice precisely the same effects: the period that the pendulum would take to swing from left to right and the time it took for the ball to hit the floor after you had dropped it would all be identical.

Interesting, eh? No matter what experiment you conduct, there is no logical way for you to tell whether you are in a gravitational field or accelerating in a region of space that is completely free from gravity. This was the hint Einstein needed to make an audacious leap of imagination: maybe gravity isn't a force at all, in the usual sense; maybe it's due to a weird type of accelerated motion. What followed, of course, was a world of mathematical pain, but when Einstein emerged out the other side, he had replaced gravity with a new concept: spacetime.

FOUR-WAY STRETCH

Now you and I both know that time and space are two separate things. After all, if I tell you to meet me at my flat, you need three space coordinates and one of time: the latitude and longitude for your satnav, for example, together with the floor that I live on, and 'just in time for *Saturday Night Live*'. We say in this way that space is three dimensional and time is one dimensional. Einstein abandoned this idea, saying that space and time are mixed, and the amount of mixing can depend on the speed that you are travelling. When you move slowly, space and time appear unmixed. This is the world as we humans see it. On the other hand, when you approach the speed of light, the mixing increases.

Don't worry if this appears baffling; it should, as your whole brain has been wired up to think of the world in one particular way, and Einstein is asking us to look at it from a completely unfamiliar angle. But before you go for a cool drink and a lie-down, try another of Albert's simple insights that might help you through the looking-glass. It requires a little bit of concentration, but if you stick with it you will genuinely have a feel for what we mean by spacetime.

Einstein had a knack for performing experiments without ever having to leave his armchair, and this is another great example. Imagine, like Albert, that you have a slightly dull office job, and you pass the time by staring hopefully at the town clock, just across the square from your window, willing it to strike 5 p.m. Since you are stationary relative to the clock, time and space are unmixed, and you experience each tick of the second hand in its full glory. What's happening, of course, is that beams of light are travelling from the clock face and landing on the back of your eye, where your brain forms them into an image and thereby works out the time. So far, so straight-forward.

But here's the thing. Imagine that as the clock strikes 5 p.m., you head off at the speed of light, away from the clock face. As far as you are concerned, the time on the clock will always remain the same. No light from any 'later' time, not even one second past 5 p.m., will be able to catch you up. In other words, for an observer travelling away from the town clock at the speed of light, time would appear to stand still for the clock. Are you getting the feel of it? A bored office worker sitting stationary in front of the town clock sees it ticking at the normal rate. A supercharged office worker travelling at the speed of light away from the same clock sees the time on the clock stop dead. It's not much of a leap now to see that the faster we travel with respect to a clock, the slower the time on the clock will appear to pass.

Still don't believe me? Then you might be interested to know that we have exhaustively tested this theory with incredibly accurate atomic clocks in aeroplanes, and every time it comes up trumps. Generally, here on Earth we travel at such low speeds that we don't notice the relativistic mixing of time and space, but in the satellite business, for example, where stuff in orbit tends to be winging about at high speed relative to things on the ground, these kinds of corrections are bread and butter.

SPACEY SPACETIME

With the concept of spacetime, embedded in the theory of General Relativity, we finally have a framework that can help us understand what's going on with big things like stars, planets, black holes and the Universe. Large objects, Einstein says, distort spacetime, and what we think of as gravity is really other objects responding to that distortion. The Sun, for example, warps the spacetime around it into a sort of valley that our plucky little planet finds itself trapped in, a bit like a golf ball rolling around a golf hole after a tricky putt.

Light follows the shortest path it can between two points in spacetime and, like matter, it also gets diverted by the spacetime 'valleys' that surround large gravitational objects. We can see this in a fascinating phenomenon called 'gravitational lensing', which tends to crop up when you get a black hole on the line of sight between your telescope and a distant galaxy. Rather than the galaxy being completely obscured, the curved spacetime around the black hole acts as a kind of lens, producing an effect a lot like a funfair mirror, with the galaxy showing up as a ring, an arc, or even multiple images.[14]

And finally, thanks to General Relativity, we have an answer to our puzzle of the exploding Universe. The galaxies aren't exploding outwards; instead, space itself is expanding, taking

---

14 There's a bit of a subtlety here, in that Newtonian gravity also predicts a deflection, but half of that produced by General Relativity. The British astronomer Arthur Eddington famously made a trip to Brazil in 1919 to record the solar eclipse. He showed that the light from stars that appear near to the Sun was being deflected by the amount predicted by Einstein, not by Newton.

matter with it. The Universe, it turns out, is less like an exploding bomb and more like a baking fruit cake, where the currants represent galaxies. As the fruit cake rises in the oven, the currants move apart from one another. Ignore the cake, and they appear to be exploding apart; include the cake, and you can see that they are being carried along by its expansion. Space, we are discovering, is far from 'empty'; it is the unseen driving force behind all of Creation.

So now you know. The night sky is black because the Universe had a beginning and is still expanding today. The stars are huge bonfires of atoms, blazing light: the biggest live fast and die young, exploding in supernovae that produce all ninety-two elements of the periodic table. What's left behind will pack down to either a hot diamond called a white dwarf, a supremely dense neutron star, or a stellar-mass black hole. Our own galaxy, the Milky Way, appears as just a strip in the night sky because it's a flattish spiral and we are looking at it from the inside out. Our solar system sits about halfway out from the hub and, like everything else in the galaxy, it is in orbit around an enormous black hole at the galactic centre that has a mass 4 million times that of the Sun.

As we know from General Relativity, what is really going on here is that this monster black hole at the centre of the galaxy has warped the spacetime around it into a deep well, and we, as well as all the stars and gas and dust and spiral arms around us, are in free fall through curved four-dimensional spacetime, accelerating towards it like so much bathwater swirling round a plughole. Luckily we are far enough away, and moving with enough momentum, that we will never get swallowed up; nevertheless, in a very real relativistic sense we are all falling into a supermassive black hole.

THIS IS THE END

So let's play this game. Imagine you are out in your experimental spaceship, miles from anywhere, testing whether you can tell the difference between acceleration and gravity by experiment, and rapidly coming to the conclusion that you can't. Exhausted, you decide to stick the ship on autopilot and bunk down for an hour or two. In your post-experimental fuddle, you accidentally set your ship's coordinates for the centre of the Milky Way and hit hyperdrive.

The ship wakes you with a Teasmade a couple of hours later, and you roll up the blind to see, to your intense dismay, that you are about to fall into the supermassive black hole at the centre of the galaxy that astronomers call Sagittarius A★. What happens next? What would you see? And what would a nearby rescue ship see? Well, here's our best guess, in accordance with General Relativity.

Remember how a stellar black hole would rip you apart long before you reached its event horizon? Well, it turns out that that's not the case for a truly enormous black hole like the one at the centre of the Milky Way galaxy. It's truly terrifying to think about, but when black holes get really large, it becomes perfectly possible to cross their event horizon without even noticing.

So let's pick up where we left off, with you in your spaceship, drifting towards Sagittarius A★, desperately trying to reconfigure your spaceship's coordinates so that you can get the hell out of there. For a black hole that massive, the event horizon would be roughly 30 million kilometres out from the centre. As you approached from fifty times that distance, the first thing you might see is a bright spot, where stars and galaxies out behind the black hole are being lensed in front of it by its powerful gravitation.

As you got even closer, the disc of the black hole would resolve into view, surrounded by a whirling ring of densely packed stars and galaxies. This, again, is a lensing effect, where the curving spacetime at the edge of the black hole is bending the light from all the stars around it; in some cases, you might even see the same star twice. No light is emitted from the monster, of course, so although it is very much three dimensional, it would appear extremely odd to your naked eye: like a flat black circle, almost as if part of the Universe has been 'cut out'.

You wouldn't notice anything odd, really, as you passed the event horizon. Your instrument panel might tell you that you were past the point of no return, but as far as the view from your rocket window was concerned, you would just appear to be approaching closer and closer to the surface of the black hole. The sky around the horizon would be crowded with blue-shifted stars, again because of the strange curvature of spacetime near to the hole, enabling you to see stars and galaxies that are, in reality, on the other side of it.

Your last moments would be rather undignified, because as you neared the 'singularity' at the centre of the hole where all the mass is concentrated, the immense curvature of spacetime would start to stretch your rocket – and you with it – into a thread. That's if the blue-shifted radiation raining down on you hasn't fried you first, of course. What happens next is anyone's guess; our physics doesn't really tell us what goes on inside a singularity, but I'd hazard that, for you, it's not great news. There is a theory that, for a rotating black hole, the singularity is avoidable if you can find a way into a wormhole that then spits you out in another Universe, but I wouldn't count on it. Even if you did manage to pop out in another Creation, what's to say that the laws of physics will be the same? You might find you grew an enormous Mexican moustache just before your atoms spontaneously combusted.

So what of the crew of the rescue ship, sent too late to try to stop you falling in: what would they see? Weirdly, they would never see you cross the event horizon. They would see your ship fall closer and closer to the dark surface of the black hole, then appear frozen in time, red-shifting and red-shifting, until you faded completely from view.

STUCK IN THE MIDDLE WITH YOU

And so we have it. In the previous chapter we day-tripped in the bizarre world of particle physics, courtesy of the Large Hadron Collider; now we've rocketed to the other extreme and gawped at large-scale features of the Universe such as galaxies and black holes. We've seen how the Sun and planets of our solar system formed from a collapsing glob of hydrogen and interstellar dust just under 5 billion years ago, and we've taken a stab at how our very own Earth acquired its glorious oceans. We've grappled with two extraordinarily successful theories: quantum mechanics, which best describes small stuff, and General Relativity, which does a great job for the big stuff. Now we get to explore all the fascinating stuff that happens in the middle.

When you pop to the shops, you generally don't have to worry about running into a Higgs particle or a black hole. Quantum mechanics may have been used to design every electronic device in your house, and the satnav in your car wouldn't work without our knowledge of General Relativity, but nevertheless it's perfectly possible to get along in life without understanding – or appreciating – either of them. A tragic existence, if you ask me, but possible. Our next chapter, however, is need-to-know stuff. It's time to see what science can tell us about ourselves: what drives us, where we came from and where we are heading. We are about to meet what is arguably science's most successful theory: evolution. And, as we shall shortly see, the astonishing truth is that you and I are nothing more than pimped-up fish mutants in the process of adapting to feed on the milk of another species . . .

CHAPTER 4

MANKIND AND OTHER FOSSILS

LOVING THE ALIEN

I don't know if you've ever seen a scan of a human foetus in the womb, but let me tell you, it's a discomfiting sight. There you are, seeking out the irresistibly adorable features of a future loved one, and what you're confronted with – within the first month or two, anyway – looks at various times like a tubular piece of plankton, an embryo trout, a wannabe salamander and a miniature Jonathan Merrick. After about twelve weeks, of course, nature plays some sort of origami trick and, by refolding a membrane here and reworking a gill slit there, something appears that looks vaguely human, but you can't help feeling it was a close call. Humanity is a veil drawn over something very inhuman indeed.

What we are glimpsing, of course, with baby-to-be photos,

is the basic genetic similarity of all creatures with a backbone. The embryos of codfish, bullfrogs, kestrels and game-show hosts all start life looking remarkably similar, because, incredible as it seems, they all sprang from a common ancestor. What's more, biology shows us that not only vertebrates but all life on Earth is ultimately related: bacteria are kin to butterflies, and William the Conqueror is a distant cousin of the nematode worm. Exactly how that is possible, and just what it is that the vast menagerie of Earth's species have in common, will be merely one of the highlights of the following pages.

There can be few questions more enticing than 'What is the secret of life?' but that is the age-old riddle that modern-day science has come tantalisingly close to answering. Once we have grasped the simple mechanism that drives all Creation, written time and time again in our planet's oldest rocks, we can start to answer some other equally fascinating questions: what, exactly, are human beings, where did we come from and where are we headed?

The mechanism I am talking about, of course, is the one and only theory of evolution, arguably the crowning glory of the whole of science. General Relativity may be impressive in its scale, and quantum mechanics may be intriguing in its sheer weirdness, but for day-in, day-out scientific success, evolution is extremely hard to beat. If one of the qualities we seek from good science is the ability to slice through a whole tangle of seemingly unrelated stuff with a single guiding principle, then evolution is a very keen tool indeed. Because evolution, incredible as it seems, manages to do nothing less than explain the extraordinary variety and complexity of life on Earth, starting with only a few overgrown sugar molecules. If you have trouble believing that human beings share a recent common ancestor with chimpanzees, or that a series of chance mutations could eventually produce something as complicated as, say, a person who believes in Intelligent Design, then this is the chapter for you.

THEORIES, HYPOTHESES AND HUNCHES

A 'theory', of course, as far as science is concerned, is a very different thing from a theory in everyday life. For example, I have a theory that the song 'From Russia with Love' was partly inspired by 'How Much is that Doggy in the Window?' based on an observation that the notes for the lyric 'I've trav-elled the world' matches that of 'How much is that do-', but very little else. I am conducting no research, planning no future experiments and am not preparing a peer-reviewed scientific paper for publication any time soon. A scientist would say that my 'Doggy in the Window' theory is really a 'hypothesis': a piece of unfinished business, an informed guess that requires something in the way of evidence to back it up.

On the other hand, when scientists talk about the 'theory of evolution', what they mean is that the ideas within it have been tried and tested to a point where no reasonable person need call them into doubt. Unlike physics, and to a certain extent chemistry, biology is a subject where it is hard to be exact; living organisms are far too complex for that, and there are a great many experiments that it is quite simply unethical to carry out.[1] So it's perhaps even more remarkable that one of the most powerful and complete scientific theories emerged in a discipline that is notoriously hard to simplify. In order for us to appreciate evolution in all its glory, let's start by reminding ourselves of the complexity of the puzzle it set out to solve.

THE ORIGIN OF SPECIES

Let's imagine for a moment that you are the Man in the Moon, looking back at the Earth, playing a kind of grandmother's footsteps game with terrestrial life. For aeons at a time, you sit with your back turned towards the developing planet, turning round only now and then to check whether anything's doing.

At the start of the game, you watch as the Earth forms, a black smouldering ball of molten rock with a dull red glow. Its surface cools to form a crust. Volcanic eruptions force a huge amount of hot gas up to the surface and out of the crust, which is then captured by gravity, forming the Earth's very first atmosphere. As you look out across the solar system, some-thing similar is happening to Earth's neighbouring planets, Venus and Mars; little Mercury, being not much bigger than our own Moon, is too small for gravity to be able to do the job and its volcanoes simply vent directly into space.

---

1 This, of course, is what keeps a lot of homeopaths and astrologers in business. It would just be wrong to stuff a Sagittarian full of arnica when no one has the faintest idea what the side effects might be.

You look away for a billion years. When you look back, you see the Earth's surface has long cooled and, apart from a few scattered volcanic islands, is now covered with greenish, iron-rich water. You have of course missed the Late Heavy Bombardment, when soggy asteroids pounded the planet. For reasons not entirely understood, although the Sun is less bright than it is in the present day, the planet as a whole is much warmer and there are no signs of ice at either pole.[2] The whole place is like a warm, oxygen-free bath and, crucially, single-celled life is present in the form of bacteria, doing very nicely, thank you, by absorbing hydrogen and carbon dioxide and giving off methane.

Again you look out across the solar system. Venus is barren; it is too close to the Sun for liquid water to form. But on Mars, water appears to have come and gone, leaving behind the famous empty canals. The puzzle of how Mars came to lose its water is still with us. One suggestion is that an asteroid collision switched off its magnetic field, meaning it was no longer protected from the charged particles of the solar wind, and its atmosphere was blasted off into deep space. Either way, alone in the solar system, the Earth is looking like the only safe harbour for life as we know it.

2 We believe that, compared with the present day, the Sun was 20–30 per cent less bright. So if there was less energy arriving from the Sun than there is today, why was the planet warmer? One possible explanation is that the atmosphere was full of volcanic greenhouse gases, which, as we shall see Chapter 7 on global warming, trap the Sun's energy. Another is simply that the Earth was less reflective than it is today, as water is darker than land, and the vast ocean simply absorbed more of the Sun's radiation than the present-day land-and-sea combo. Yet another is that it may require the existence of continental land to get the Earth's surface water high enough – and therefore cold enough – to be able to freeze and start to form an ice sheet. The thorny question of how the Earth's climate was so much warmer, even though the Sun was less bright, is known as the Faint Sun Paradox, and was first mooted by the inspirational American cosmologist Carl Sagan.

You look away for another billion years, and then you look back. It's now 2 billion years since the Earth formed. The changes are slight, but significant. There's more land, probably about a quarter of what we see today, and the water's surface is now home to huge blooms of single-celled blue–green algae, which are merrily hoovering up carbon dioxide and producing the Earth's first lungfuls of free oxygen. As of yet, little of this oxygen is making it into the atmosphere.[3] Instead, it's busy reacting with anything and everything it can get its hands on, such as rusting the iron in the world's oceans into iron oxide and reacting with atmospheric methane to produce the planet's first ozone.

If you look away for another billion years and look back again, life has inched another important step forward. For the first time, complex single-celled creatures are floating around whose structure is very similar to the individual cells of present-day plants and animals. The oceans are now blue, as free oxygen in the atmosphere has reacted with all the iron in the world's oceans; the amount of oxygen in the atmosphere is around a fifth of that we see today.

Look away for a further billion years, then look back, and you'll finally see the vague trappings of a planet you recognise, and with it the first stirrings of multicellular life. There are lichens growing on land and jellyfish bobbing about in the sea. The continents are all squashed together, having just broken up from one enormous supercontinent, Rodinia, and there are ice caps at the poles. Oxygen in the atmosphere is nearing present-

3 Or, indeed, into anything's lungs, as no creatures have evolved them yet.

day levels and there is a fully protective ozone layer over the entire planet that is filtering out the Sun's harmful ultraviolet rays.

Finally, you look back in the present day. And here's the crazy bit. You see the Earth exactly as it is today, the continents all where they should be, and every corner of the globe teeming with life. Almost every species of life on Earth just happened, and you missed it. You missed the first trilobites, the first fish, the first amphibians, the first reptiles, the first dinosaurs, the first mammals, the first apes and the first humans. They all checked in – and, in most cases, checked out again – in the last half-billion years of Earth's 4.5-billion-year lifetime. What's more, one of them – *Homo sapiens sapiens* – developed language, culture and technology, and then had the remarkable good sense to go and buy this book. What theory could possibly account for all this?

THE SURVIVAL OF THE WHAT?

In layman's terms, evolution is simply the idea that species can change.[4] The world may now be populated with kestrels, bull-frogs, human beings and the like, but it was once populated with something completely different. And what, you may ask, brings about change in species? Natural selection, that's what. In any given population of organisms, be they bacteria, marmosets, or

---

4 And what, you may ask, is a species? Two animals are of different species if they aren't able to produce viable offspring. Thus a cow and a pig are different species, whereas a Chihuahua and a Great Dane are not. Though a happy ending for a Chihuahua and a Great Dane is clearly a big ask, for practical reasons.

farming villages in the Dolomite mountains, over sufficient generations any trait that tends to bring about more offspring will become more and more widespread. Squirt a colony of bacteria with disinfectant, and those that are most resistant will prosper. As will those marmosets most capable of hanging on to the tree canopy in a high wind, and men with long noses in the aforementioned Dolomite village if that's what their womenfolk decide is attractive.

All very well, you might say, but an Italian village full of men with long noses is not exactly a new species. And you would be right. But if an isolated population survives through sufficient generations, enough change can take place through natural selection that a whole new species emerges. The giraffe and the okapi, for example, are descended from a tall deer-like mammal that roamed Africa over 5 million years ago. The fact that giraffes have long necks and okapis have short necks leads one to suspect that two groups of their ancestors found themselves isolated in slightly different circumstances, and that maybe one of those circumstances involved a lot of delicious acacia trees and not that much tasty stuff on the ground.[5] We weren't there, so we'll never really know. The point is that

5 To be really accurate about the way Darwin's theory of natural selection works, the variation comes first, followed by the selection. So the idea is that giraffes with long necks are able to reach the leaves on acacia trees, and therefore have more offspring than giraffes with short necks. One of the prevalent theories of evolution before Darwin, called Lamarckism after the French zoologist Jean Lamarck, proposed that – if you get my drift – the tall trees came first, then the long necks. In other words, straining to reach high leaves lengthened the necks of ancestor giraffes, and their offspring inherited that trait. To put it in a nutshell, Lamarck saw evolution as full of purpose, whereas Darwin saw it as blind chance. Lamarck's view might make us all sleep a bit easier at night – after all, it would be great if life had a point – but Darwin's angle is the one supported by the facts.

giraffes and okapis have changed so much, compared to their tall deer-like ancestors, that they are now different species.

I HATE THIS THING I DON'T UNDERSTAND

It's a testament to the success of Charles Darwin's theory that the word 'evolution' has become part of our everyday vocabulary, though, it has to be said, we don't often use the word correctly. We might say, for example, that over the last twenty years the England cricket team has 'evolved' into a lean, efficient fighting machine, or that the Tesco chain has 'evolved' into the UK's number one supermarket. What we mean, usually, is that the thing we are talking about has improved slowly and steadily over time. Yet evolution as Darwin meant it doesn't imply that organisms 'improve' in any objective sense; shoals of sharp-eyed fish trapped in pitch-black underground caves will take only a few generations to become sightless, for example, and forty-odd thousand years in rubbish weather has robbed Europeans of their ability to tan in the Sun, or at least it has in my case. Evolution doesn't have to change things for the better, and it doesn't have to change them slowly, either; new generations of viruses pop up that have found a way to sidestep our best vaccines, for instance, and on my tropical travels I seem to constantly come across well-adapted mosquitoes that are not evenly remotely put off by my multiple layers of DEET.

We talk, too, about 'survival of the fittest', in the context of companies going to the wall in a recession, or an antelope running from a hungry cheetah: the general idea being that life is a battle for survival, with the winner as the one that is still standing at the end of the fight. But weirdly, again, this isn't the exact sense that 'survival of the fittest' has in biology. When biologists talk about an organism's 'fitness', they aren't talking about the size of its muscles, but about the number of offspring it manages to produce. If you make a lot of babies, you are fit. If you make none, you lack fitness.[6] All that matters for an organism is how many copies it makes of itself. Which is to say that, as far as evolution is concerned, the question isn't whether the antelope escapes the cheetah; the question is whether the antelope manages to produce lots of little antelopes before it either escapes or gets ripped to pieces.

UNINTELLIGENT DESIGN

Evolution – big, bold, beautiful evolution – is a deceptively simple theory with an extraordinary power to explain the enormous variety of life we see around us. It is a magical thread that can pull together all the evidence of geology, palaeontology and genetics into a coherent glittering whole. And yet we so often misrepresent it. And that's those of us who are well disposed towards science: broad-minded individuals who

6 And if you think about it, in the case of men who donate to sperm banks, it is perfectly possible to be extremely fit in the evolutionary sense – in other words, to have lots of offspring – and yet to be stone dead.

regularly entertain the idea that there are such things as genes and fossils and radioactive dating of rock samples, and who are happy for such things to be taught in our schools. But the lazy way that we sometimes approach Darwin's theory is nothing compared to the kicking it gets from the religious right. The fossil record is incomplete, they say, and there is no proof of a link between apes and humans. Organisms appear so wondrously adapted to their surroundings that they must have had a creator. Evolution, in their view, is only an opinion and ought to be taught as such.

So how exactly did our understanding of evolution come about and what is the evidence for it? It's time to get to the facts of the matter. And no one could wish for a more painstaking fieldworker – or, indeed, a more visionary theorist – than the extraordinary Charles Robert Darwin. And, as we shall see, a chance encounter with a chaffinch was the vital clue that put this master detective on to the scent of evolution by natural selection.

A CHIP OFF THE OLD BLOCK

There is considerable evidence that intelligence is a heritable trait. If it is, then that might go some way to explaining Charles Darwin's genius. He was, after all, a product of two extraordinary Staffordshire families: a grandson of the celebrated potter Josiah Wedgwood on his mother's side, and a grandson of the eminent physician and naturalist Erasmus

Darwin on his father's. Wedgwood used the scientific method to revolutionise pottery, making a fortune in the process and indirectly ensuring that grandson Charles didn't have to work for a living. The inheritance Erasmus Darwin bequeathed was no less significant: he was one of the first men to contemplate the idea we now know as evolution.

Erasmus was a man of great appetite: for women, pies and ideas. He prescribed sex as a cure for hypochondria and, if his fourteen children are anything to go by, was more than happy to take his own medicine. His gourmandising was legendary, and in his later years he is rumoured to have cut a semicircle out of his dining table in order to accommodate his belly. A poet, botanist and inventor, he was so fond of hypothesising that Samuel Taylor Coleridge coined the term 'darwinising' to mean speculating wildly. And yet in one extraordinarily pre-scient phrase in his *Laws of Organic Life*, published in 1794, he declared: 'Shall we conjecture that one and the same kind of living filament is and has been the cause of all organic life?'

To be fair, there were a number of other people conjectur-ing the same thing around the same time. The first fossils had been identified as such in 1666,[7] and back in the mid-1700s Georges-Louis Leclerc, the Comte de Buffon, had proposed that the similarity between many of Earth's species could be explained by the existence of a common ancestor. But it has to be significant for the young Darwin's developing interest in the natural world that his grandfather had a toehold on the edifice of evolution. Charles's copies of his grandfather's books

---

7 Sharks' teeth, in case you're interested.

were heavily annotated, and *The Origin of Species*[8] mentions many of the same ideas, even if an older, wiser Darwin commented that, 'On reading [Erasmus Darwin's *Laws of Organic Life*] a second time after an interval of ten or fifteen years, I was much disappointed; the proportion of speculation being so large to the facts given.'

The young Darwin was an avid collector, and no doubt it was the urge to chronicle nature rather than interrogate it that propelled him to take the voyage that was to make his name. Following the Napoleonic Wars, the British Empire was blossoming and keen to establish trade with the New World. Our hero pulled a few strings and managed to get himself hired as naturalist-in-residence on the *Beagle*, one of several survey ships being dispatched by the Royal Navy to chart the coast of South America. Darwin, diligent as ever, accumulated thousands of new plant and animal specimens, and his journal of the trip both established his reputation as a naturalist and became a best-seller.

Curiously, the find that would turn out to be pivotal was one of those to which Darwin had paid least attention on the voyage: a motley collection of songbirds from the Galapagos Islands. Back in London, Darwin and a prominent ornithologist identified them to be thirteen species of finch – an extremely peculiar result as Darwin knew of only one species of finch anywhere near the Galapagos Islands, and that was on

---

8 For the pedants, it's not *Origin of the Species*, it's *The Origin of Species*, or, to be even more pedantic, it's *On the Origin of Species by Means of Natural Selection*. And while we're at it, it's Stephen Hawking, not Hawkings, and it's 'Follow me, follow, down to the hollow' not 'Follow me, follow *me*, down to the hollow' in Flanders and Swan's classic 'Hippopotamus Song'.

the South American coast. It was then that our hero had one of the most important thoughts in the whole of science. What if a pair of South American finches had been blown on the wind to the Galapagos? If they were, then maybe this baker's dozen of new species were their descendants. He noted that the Galapagan species all had different-shaped beaks; could it be that they had adapted to different food sources, becoming so unlike one another that they were eventually incapable of producing offspring? The *Beagle* returned from its five-year voyage in October 1836. In the summer of 1837, Darwin took up his notebook and sketched his first evolutionary tree.

PUBLISH AND BE DAMNED

A reasonable person might think that, at this point, having stumbled across the defining theory of modern biology, Charles would book an evening at the Royal Society, spill the beans and lap up the glory on the lecture circuit. But, perhaps understandably, given the religious outrage he feared he might face, Darwin spent well over twenty years finding something better to do than publish on natural selection. Mostly that thing was geology, upon which he became a great authority, but there were also tomes on living and fossilised barnacles, which, if you ask me, could have waited. In a private letter to his best friend the botanist John Dalton Hooker, dated 11 January 1844, he declared almost in passing that, finally, he was preparing a 'very presumptuous work', expressing his belief that species were capable of changing, and that the very thought was like 'confessing a murder'.

A further decade and a half passed, and no dice. So when a letter from the young English naturalist Alfred Russel Wallace arrived at Darwin's home in 1858, containing a scientific paper in which he outlined his own, independently arrived-at theory of evolution by natural selection, one can only imagine it really spoiled the old man's day. Charles had nothing ready for publication, yet had embarrassingly copious notes, letters and researches that showed in painful detail that he had got there first, as well as his half-finished manuscript of *On the Origin of Species*, hinted at in the letter to Hooker. In the event, Hooker and others brokered a deal, and both Wallace's paper and an extract from *Origin* were published together. The world's most revolutionary scientific theory was finally out in the open.

A ROCK AND A HARD PLACE

Some rather unexciting-looking brown finches from the Galapagos may have been enough to convince Charles Darwin that life on Earth evolved by chance, but what evidence has been bagged and tagged since? The truth is, there's so much that it's hard to know where to start. In Darwin's time there were relatively few fossils and there was little to go on, other than observations of living organisms, which is why his expeditions to far-flung corners of the globe were so formative. In the last 150 years, however, biological science has come a very long way indeed. And, as we shall see, it's no accident that Darwin had a day job as a geologist, because the lumps of hard stuff lying all around us provide the perfect window with which to view the first faltering steps of life on Earth. It's time to talk about plate tectonics, the rock cycle and the fossil record.

WHAT GOES AROUND COMES AROUND

It's a common observation that all things go in cycles. Beards, for example, are currently all the rage in London for the first time since the 1970s though it's taken me so long to grow mine that the fashion is already moving on. In the world of science, you've probably already heard of the water cycle, and if you haven't, I can promise you that you know what it is anyway: water evaporates from the sea to form clouds, the clouds then rain onto land, the rainwater forms rivers, and the rivers run back to the sea to start the whole merry-go-round all over again.

There's another cycle you may have already heard of, too: the carbon cycle. Bizarre as it seems, it is quite likely that the carbon atoms that make up your body have already been in someone – or something – else's body before you. For example, a carbon dioxide molecule in the air might one afternoon be used in photosynthesis by a dandelion, where it is combined with water to make a sugar. A fox then eats said dandelion,[9] digesting the sugar molecule, thereby releasing energy so that

9 By mistake. Dandelions do not form part of the natural diet of a fox.

it can whisk its handsome tail and, in the process, converting the sugar molecule into carbon dioxide which it then exhales back into the air.

There are huge amounts of carbon dioxide stored all over the planet in such places as forests, sea water and the atmosphere, and they constantly absorb and release carbon in a never-ending cycle. Take a forest, for instance. On a sunny day, its leaves will merrily suck up carbon dioxide for photosynthesis and therefore remove it from the atmosphere. But if that forest were to catch fire, it would release lots of that carbon into the air in the form of carbon dioxide. Generally speaking, these different sources and sinks come into balance with each other, and the amount of carbon dioxide in the air stays more or less constant.

The carbon cycle, of course, has been in the news because of global warming. The main point to grasp is that there is also a huge amount of carbon stored underground in the form of fossil fuels like coal, oil and gas, which generally sits on the sidelines while the forests and sea water and dandelion-eating foxes do their thing. When we burn a lump of coal, we are adding more carbon to a system that is already in balance, just as if we shipped over a starfreighter full of water from Uranus and tipped it into the Atlantic Ocean. Add more water to the planet, and the sea level goes up. Add more carbon, by burning fossil fuels, and the amount of carbon in the atmosphere goes up. As we shall see in Chapter 7, we have kept a good record of the amount of carbon dioxide in the air since the 1950s, and it shows a sure and steady increase. Whether or not that is something we should all be worried about is the business of a later chapter, but for now the main points to grasp are that (a) carbon is essential to the vast majority of life on our planet and (b) there are cycles that take a lot longer to make a full circuit than the average molecule of water does. And as far as evolution goes, the most important one to understand is the rock cycle.

ROCK REVOLUTION

We think of rock as something unchanging: a foundation stone is there for ever; a marble statue preserves its subject for eternity. But the truth is that rock is far from permanent. Like everything else on Earth – water, carbon, life forms and formats for TV shows – what goes around comes around. It can take only days for a molecule of water to follow the water cycle, whereas the equivalent cycle for a rock takes around 150 million years, but it is a cycle nonetheless. That's an enormous amount of time on the human scale, but for our 4.6-billion-year-old planet it's small fry: enough time for at least thirty rock cycles since the Earth began.

The Earth is basically made up of four layers: a core of solid iron, surrounded by a layer of liquid iron, then a solid mantle, topped with a solid crust. The core is extremely hot, with an estimated temperature of over 5000°C. Given that iron melts at around 1500°C, you may be wondering how the core manages to stay solid. Incredibly, the answer is that it is under so much pressure that it simply cannot melt. The same goes for the mantle. At the top, where it meets the crust, it's up to 900°C; at the bottom, it's greater than 4000°C. Most rocks melt at 600–1200°C, so you'd expect it to be molten, but, as with the core, the enormous pressure the mantle is under keeps it from melting. When bits of the mantle burst to the surface, of course, via something such as a volcano, the pressure is released and it very much becomes liquid rock, or lava, before cooling and forming new crust. It's this extremely hot, solid-but-plastic mantle and the cold, solid crust that are responsible for the rock cycle.

Seen in one way, the rock cycle is basically the turnover of rock between the crust and the mantle as the Earth cools. Think of the mantle like a huge bowl of porridge, cooling by convection.[10] Pockets of hot rock rise up through the mantle to the surface, where they cool and form enormous great slabs of solid crust, also known as plates. Each plate pushes out across the Earth's surface, until it collides with one of its fellow plates and it gets forced back down into the mantle once more, where it reheats, becomes plastic and the whole cycle is ready to begin again.

So now you understand why the Earth's plates are moving: they are being shoved around by enormous, incredibly slow-moving thermal convection currents. And as the continents are simply the bits of the plates that are above sea level, they are

---

10 That's the one where a liquid or gas cools by means of currents. Hotter, less dense currents rise to the surface of the liquid or gas and cooler, denser currents sink to the bottom. This is why the much misused sash window is such a great invention. The idea, of course, is that you slide down the top part of the window to release a current of hot air, and slide up the bottom part of the window so that a current of cold air can be sucked in to replace it. Just sliding up the bottom part of the sash sort of misses the point.

moving as well. The whole of continental North America, for example, sits on a socking great slab of rock called the North American Plate. And here's the interesting bit: not only does what you and I call North America sit on that plate, but so too does part of Siberia to the west, and Greenland to the east, while the whole plate is heading away from the Mid-Atlantic Ridge at a rate of about a centimetre a year.

Northern Europe sits on the Eurasian plate, which is also heading away from the mid-Atlantic, but in the opposite direction. The Mid-Atlantic Ocean Ridge is the join between the two plates; in fact, ocean ridges and trenches often occur at the boundary between tectonic plates, as do mountains, volcanoes and earthquake fault lines. The famous San Andreas fault in California, for example, is the slipping joint between the north-east-bound Pacific Plate and the south-west-bound North American Plate; the Himalayas are basically a pile-up between the north-moving Indian Plate and the Eurasian Plate; and the islands of the Lesser Antilles in the Caribbean are a chain of volcanoes at the junction of the Caribbean Plate and the southernmost portion of the North American Plate. The world map that we are introduced to at school, and which seems like such a permanent fixture in our hearts and minds, is really nothing more than a snapshot. The country you call home, the continent that contains it, and maybe even the mountains you live on, are all just a temporary feature of the 150-million-year rock cycle.

GRANITE ROCKS

Ah, I hear you say, but if the rock cycle lasts around 150 million years, surely there should be no rocks any older than that in existence, and no way of discovering anything about the Earth and its life forms at any time pre-dating the Jurassic. And you would be right, but for one wondrous thing. Granite.

Geology, you see, is basically a game of two rocks, basalt and granite. Both are what are called *igneous* rocks (from the Latin for fire, *ignis*), meaning that they are created when molten rock cools. When molten rock from the mantle cools quickly, as it does when it wells up between two spreading tectonic plates on the seabed, you get dark, fine-grained, dense basalt. When it cools slowly, you get light, coarse-grained, less-dense granite. Granite tends to form when hot rock from the mantle wells up into the crust, but fails to reach the surface. It therefore often spends the first part of its life underground, emerging only once the rocks above it have weathered away.

So, what's so special about granite? Well, you may think of it as being heavy, but in geological terms it is as light as a feather. The very first rocks on the planet would have all been basalt, formed as they were from molten rock that had cooled rapidly in the air or the ocean. But, fairly soon after, those first rocks would have weathered away to reveal newly formed granite, and granite is much lighter than basalt. When two plates meet, one made of granite and one of basalt, the basalt tends to end up underneath, in the same way as, in a car crash, a heavy Volkswagen ends up under a featherweight Reliant

Robin. The basalt, in other words, is forced back down into the mantle and recycled, whereas the granite is so light that it ends up floating on top like an air-bed in a swimming pool. Granite, clever stuff, frequently bypasses the destructive phase of the rock cycle.

Because it often avoids being recombined with the mantle, granite has a chance to accumulate. In fact, that's what continents are really: huge great lumps of granite. The seabed, on the other hand, tends to be made of basalt. What all of this means, of course, is that a good chunk of the granite made since the formation of the Earth is still knocking around. As discussed earlier, we haven't found any of it that dates right back to the very beginning of the Earth, but there are some large lumps, or *cratons*, that come close.[11] And if you're really lucky, large lumps of granite can act as a sort of life-raft for other rocks. At the time of writing, the oldest known rock on Earth is actually a piece of pinkish basalt, trapped in a large lump of granite in the Nuvvuagittuq Greenstone Belt on the eastern coast of the Hudson Bay in northern Quebec and thought to be 4.3 billion years old. But as far as evolution is concerned, there are other, even more exciting rocks that – thanks to the buoyancy of good old granite – have found their way to us across countless oceans of time. And ridiculous as it seems, these mini-miracles have fully preserved the three-dimensional outline of the very first life forms that lived on Earth. It's time to meet one of the true wonders of the planet: fossils.

11 The Kaapvaal *craton* in South Africa, for example, is thought to be some 3.6 billion years old.

THOSE BONES, THOSE BONES

Let's just consider the seemingly overwhelming odds against any animal becoming a fossil. In fact, what are the odds of you, or anyone you know, being preserved in rock for, say, half a billion years and then being discovered by some future life form?

Well, for a start, you can't just keel over and die anywhere. You had better do it in a river, lake or seabed, and your dead body had better be undisturbed by predators and covered rapidly by fine sediment before your bones can rot. Then the water you have been buried in needs to contain plenty of dissolved minerals (silicon dioxide, calcium carbonate, iron sulphide and the like) so that it can seep through the sediment and replace the organic material of your bones with some good, old-fashioned, hard-wearing, inorganic chemical compounds.[12] Then the whole contraption needs to be covered by many more layers of sediment over a period of millions of years, so that the whole caper can be compressed into rock.

Eventually, your bones will be preserved in mineralised, rocky form in layers of sedimentary rock. That sedimentary rock now has to avoid being reheated in the mantle as the rock

---

12 Chemistry is basically divided into two: stuff involving carbon, which is called organic chemistry; and everything else, which is called inorganic chemistry. So sugar, for example, which has a molecule with a carbon backbone, is an organic compound, and copper sulphate – which forms those beautiful blue crystals – is an inorganic compound. Generally speaking, life forms use carbon chemistry. In fact, sugar molecules, as we shall see in the next chapter, turn out to be some of the most important players in the whole of evolution.

cycle does its thing; 500 million years will involve something like three to four entire rock cycles. As we've just seen, the best bet is for the sedimentary rock to hitch a ride on a large lump of granite, and that lump of granite has to be lucky enough to avoid getting forced back down into the mantle.

If all that sounds like a long shot, consider this: you may have been fossilised and avoided an early bath in the mantle, but you haven't been discovered yet. That's a whole other set of odds. For you to be discovered, the upper layers of the sedimentary rock that you sit in have to be weathered away at just the right rate to leave your bones exposed at exactly the moment that a future fossil hunter arrives, in 500 million years' time. And he will need to be really on the ball, with an expert team at his disposal; after all, maybe there's just one of your toenails showing among a treasure trove of fossilised leaves and whelk shells. Furthermore, the chances are that the site is a long way away from civilisation – how else could it have been exposed by the elements? – so he will also have to have some fairly hefty funding in place and an almost obsessive drive to uncover the secrets of the past with little public recognition or reward. Such men are rare. In short, good luck.

ONE OF YOUR ANCESTORS WAS A FISH

If you ask me, it's utterly remarkable that even one fossil has been found, let alone the countless thousands that sit in museums and private collections all over the world, a silent stone jungle of extraordinary plants and creatures from every imaginable age and corner of the Earth. Not surprisingly, given that soft tissue rots so quickly, we find very few that date from earlier than 540 million years ago, when creatures first evolved hard body parts. That period of the Earth's history is called the Cambrian, and the sudden appearance of fossils in sedimentary rock of that age is known as the Cambrian Explosion. Nevertheless, we have some fossil traces of earlier, soft-bodied creatures and of the single-celled animals that preceded them; that, together with the evidence of geology, is how we are able to build the picture that I described with the grandmother's footsteps game at the beginning of the chapter. But if you asked me to name just one piece of fossil evidence that makes evolution a fact that no reasonable person could doubt, we need look no further than a rather peculiar 370-million-year-old fish.

A crucial feature of a good scientific theory is that it makes predictions that you can test. Evolution has it that natural selection acts upon the varied traits of a given population of organisms, such as the descendants of two plain-looking finches in the Galapagos Islands, eventually producing new species. If that is the case, and we think that, for example, amphibians evolved from fish, then some intermediate species must have existed that was somewhere in between the two. In other words, we are looking for a fish with legs.

Given what we know about how fossils are formed, we have a reasonable chance of finding this creature. Fish, after all, live in rivers, as do frogs, and rivers have sediments, so there is a fighting chance that our fish with legs ended up as a fossil. If we know the age of the rocks in which we find the first fish, and the age of the rocks in which we find the first amphibian, then we'd expect to find the first leggy fish in rocks aged somewhere between the two. All we need to do is pinpoint a place on Earth – most likely remote – where sedimentary rocks of that age are exposed, and go fossil hunting until we discover something. And in 1999 that was what an extremely dedicated, well-funded palaeontologist called Neil Shubin and his team set out to do.

At that time, fossilised fish had been found in rocks 380 million years old, and fossilised amphibians in rocks 365 million years old, so Shubin and his team quite rightly went on the hunt for sedimentary rocks formed in large, shallow river beds that were 370 million years old. After doing their geology homework, his team tracked down some sedimentary rocks on Ellesmere Island in Ninavut in northern Canada that really seemed to fit the bill. A few fruitless expeditions followed, in which the team was helicoptered onto the ice and left for months, finding virtually nothing. However, on a return visit in 2004, a junior member of the team failed to return to camp by curfew, and eventually staggered up way after dark with his pockets laden with fossils. Together, the group managed to relocate the sweet spot and, after a lot more hard work, dug up nothing less than a creature they named Tiktaalik, which – as you can see – could only go by the description of a fish with legs.

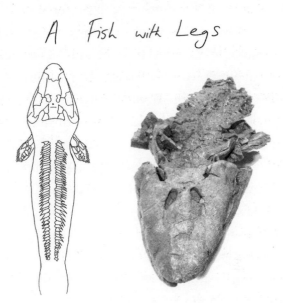

A Fish with Legs

LATE TO THE PARTY

As we saw in our grandmother's footsteps game, evolution has packed an awful lot into the last half a billion years, as compared with the preceding 4 billion. And as far as mankind is concerned, we really have turned up very late to the party indeed. Despite what you may have seen at the movies, we have never rubbed shoulders with dinosaurs; their last hurrah was 65 million years ago, at the end of the Cretaceous period.[13] Dinosaurs had truly 'ruled the Earth' for the preceding 160

13 The name Cretaceous, by the way, comes from the Latin *creta*, meaning chalk. At this time, the Earth was warm with shallow, carbonated seas that hosted an army of single-celled algae called coccolithophores. When these algae died, they left behind minute shells made of calcite, also known as calcium carbonate. Millions of years later, that calcite was compressed into a sedimentary rock: chalk.

million years, but, as you may recall from the last chapter on black holes, their good fortune came to an abrupt end with the asteroid impact that created the Chicxulub crater in Mexico.

Once dinosaurs were out of the way, mammals and birds were quick to fill the gap. Even so, primates didn't evolve from their squirrelish, shrew-like ancestors until around 58 million years ago. And primates didn't evolve into the great apes, or hominids, the family that humans belong to, until roughly 15 million years back. At least, that's as far as we can tell. The first hominids, after all, lived in trees, not rivers, and the chances of them ending up as fossils are therefore pretty slim. The ancestor of the orangutan evolved first, then that of the gorilla, and the last common ancestor of chimpanzees and humans lived only about 6 million years ago.[14]

Once chimpanzees had gone their separate way within the hominid family, the last remaining line was that of our human ancestors, the hominins. The earliest complete hominin skeleton we have was found in Ethiopia and has been dated to some 4.4 million years old. She's called Ardi and she's worth a Google. She looks nothing like a chimp, except in her general size, although to be fair she doesn't look that much like a

---

14 It's kind of worth pointing out a sad fact of evolution: most species die out. We know that we are related to gorillas, for example, but we have no idea what our common ancestor would have looked like because they quite simply didn't stick around. If this seems odd, just think of telephones. Look at an iPhone and a regular landline phone. They look nothing like one another, but they share a common ancestor. That common ancestor is no longer around, because people tired of holding a mouthpiece in one hand and an earpiece in the other, and shouting 'Knightsbridge 215' and suchlike while dancing the Charleston. I suppose the point I'm trying to make is that just because humans don't look like chimps doesn't mean we don't have a common ancestor. And that common ancestor might very well have looked nothing like either of us.

human either. She walked on two legs, but she also had an opposable big toe, suggesting that she was no stranger to the odd tree. Ardi isn't the 'missing link' that you hear talk of that is the common ancestor of humans and chimpanzees; by the time she was alive, the chimpanzee and human lines had long since parted company. But clearly she was no longer only swinging through the trees so beloved of her fellow great apes, but also commuting two-legged along on the forest floor.

OUT OF AFRICA

So around 15 million years ago, the hominid family emerged in Africa, and some 6 million years ago our own walking, talking hominin line separated from that of chimpanzees. To be precise, of course, our ancestors probably weren't walking and talking 6 million years ago, which leads us neatly to the burning question: just when and where did they start to develop language, fire, art and Grand Theft Auto?

The answers aren't easy to come by, and part of the problem is the very nature of fossils. Fossils, as we know, are extraordinarily rare, and fossils of creatures that don't spend most of their day flopping around on a beach or a river bed are even rarer. The fragments of hominin remains that we tend to find – a few teeth here, an arrowhead there – are frustratingly hard to come by. Nevertheless, there are a few headlines that all palaeontologists pretty much agree on, and they offer a fascinating glimpse of our birthright.

First, Africa is the place where it all started. The 4.4-million-year-old Ardi is followed by the 3.2-million-year-old Lucy, also from Ethiopia but of the genus *Australopithecus*, whereas Ardi belongs to *Ardipithecus*.[15] Lucy lacks the opposable toe of Ardi, and is thought to have been less of a tree dweller; the point to note, though, is that neither of them was any great shakes in the cranial department. Whatever caused our earliest ancestors to abandon the trees, it appears not to have been a sudden surge in mental ability, at least if skull capacity is anything to go by.

After the likes of *Ardipithecus* and *Australopithecus*, at around 2.3–2.4 million years ago, we start to find fossils of our own genus, *Homo*. Our family tree is said at this point to be 'bushy', meaning there doesn't really appear to be one long clean line of inheritance, but rather an untidy profusion of different *Homo* species, some of which could be our direct ancestors while others are evolutionary dead ends. One thing is plain, though: the new arrivals have bigger brains. In rough chronological order, we find older species such as *Homo habilis*, *Homo ergaster* and *Homo erectus*, and more recent ones, such as *Homo heidelbergensis*, *Homo neanderthalensis* and our own *Homo sapiens*.

So what could have been the environmental pressure that selected hominins with larger brains? We know from dark and light bands in sedimentary rock from the River Nile that,

---

15 You may be wondering what a genus is. Binomial nomenclature, as it's called, was invented by the Swedish botanist Carolus Linnaeus, and it works like this: the first of the two names of an organism denotes its generic name, or genus; the second describes its species. The genus is always written with a capital, and the species never is. So we have *Tyrannosaurus rex*, often abbreviated to *T. rex*, and *Escherichia coli*, often shorted to *E. coli*. To be really formal, Ardi is *Ardipithecus ramidus*, and Lucy is *Australopithecus afarensis*.

around 3 million years ago, the climate in Africa began to swing between wet and dry periods; in fact, the whole of hominin history has seen a gradual increase in climate variability, coming to a head around 2.8 million years ago with the dawning of the present Ice Age. The general trend has been from warmer to cooler, but it's the fluctuation around that trend that many scientists believe has played an important role in human evolution. After all, big brains are good at coping with change, and there does seem to be a link between periods when the climate has altered rapidly and increases in our ancestors' brain size. Climate change may not be such a good thing today – particularly if you live in the Maldives – but for humanity in general, it may have been the making of us.

Finally, we also know that the *Homo* genus loved to travel. *Homo erectus* was the first to leave Africa, spreading out into Asia and then Europe around 1.5 million years ago. *Homo heidelbergensis* journeyed to Europe around half a million years ago, and *Homo neanderthalensis* – better known to you and me as the Neanderthals – made the trip some 400,000 years back.[16] And then, 200,000 years ago, modern humans emerged in Africa.

But though our knowledge of early *Homo* species is sketchy, when it comes to modern humans we more than make up for it. One well-supported hypothesis has it that roughly 100,000 years ago a group of *Homo sapiens* left Africa and crossed the

---

16 I should mention that no Neanderthal fossils have actually been found in Africa, so it's possibly stretching it a bit to say that they definitely originated there. In fact, there is a school of thought that says the Neanderthals actually evolved in Europe from earlier humans such as *Homo heidelbergensis*.

Red Sea at the Bab-el-Mandeb strait, following much the same path as *Homo erectus* had traced out some 1.4 million years before. They then followed the coastline of the Arabian peninsula, crossing, via the Strait of Hormuz at the mouth of the Persian Gulf, into present-day Iran. They continued east, still keeping to the coast, reaching India around 80,000 years ago, and eventually arriving in Indonesia, where the eruption of Mount Toba reduced their numbers to around 10,000 in total. Slowly the population recovered, and a cold snap beginning around 75,000 years ago saw some heading south to Australia, while others headed north up the coast of China. Things then warmed up around 50,000 years ago, and some of these modern humans travelled back up the western coast of India, across the not-yet-Holy Land into Europe, where they found the place populated by cold-hardy Neanderthals around 40,000 years ago. What happened next is the subject of some debate, but suffice it to say that by 30,000 years ago there were no Neanderthals left in Europe whatsoever. *Homo sapiens* was now the only surviving line of the genus *Homo*.[17]

The rest, as they say, is pre-history. We see a surge in art and trade, as all manner of modern human groups spread inland into central Asia, back into northern Africa and down into

---

17 I think a lot of the confusion that surrounds science has to do with the reporting of hypotheses as if they are theories. Newspapers tend to be especially bad at this, which is maybe one of the reasons people lose faith in science. When it comes to the general media, it's worth taking anything billed as a 'theory' with a pinch of salt unless there's a whole series of experiments that back it up. As a rule of thumb, anything you are reading about in a newspaper will tend to be a 'hypothesis' rather than a 'theory', and any experiment will tend to be a one-off result that needs confirming. Though if you ever come across an exposé of the misappropriation of the chorus melody from 'How Much is That Doggy in the Window?' I think we both know you've stumbled across solid fact.

southern Spain. From central Asia, we populated the Arctic circle and crossed from Siberia to Alaska, beginning the population of North America. The globe froze over 20,000 years ago, which encouraged some of these North American settlers to travel down to a new home in South America. The entire planet was now populated by modern humans, and when the present warm period began around 12,000 years ago our numbers totalled some 6 million.

It was then that another significant invention of humankind took root: agriculture. Farming seems to have started in several places at once, around 10,000 years ago, and has propelled our technology, culture, art and society to develop at an extraordinary pace. We have just passed the 7-billion population mark and have inhabited just about every niche on the entire planet. We are nowhere near the most abundant animals on Earth – that glory belongs to insects, worms and plankton – but we are undoubtedly the most adaptive, communicative and curious. So curious that we have unlocked even the secrets of our own origins.

IT'S IN THE GENES

The natural scientist in you is maybe getting a little suspicious. As far as our knowledge of our ancestry goes, we seem very sketchy on the detail up to 100,000 years ago, then almost infinitely knowledgeable about everything since. All we seem to know about our ape-like ancestors such as *Australopithecus* is that they walked on two legs and weren't that bright; we aren't terribly sure where the Neanderthals first evolved – Africa or Asia – and we have only the sketchiest details of *Homo sapiens'* emergence in Africa. And yet we seem to know the name, address and postcode of every *Homo sapiens* to leave Africa in the last one hundred millennia, and the exact route they took to get there. How come? Did someone uncover a trail of hand-axes, or a Neanderthal teenage diary?

We are now getting to the meat in the sandwich of the evolution story, because our evidence for *Homo sapiens'* migration comes not just from the fossil record, but also from the cells of our own bodies. Our next jaunt is into the magical world of DNA, an extraordinary sugar molecule that holds the blueprint for life, a blueprint that – even more extraordinarily – we are able to read. And, as we shall see, DNA also holds vital clues as to where we came from and when. What's more, those clues can take us far back in time, far beyond our emergence in Africa, even beyond our common ancestor with chimps and the other great apes. In fact, it can take us right back to the beginning of life itself.

CHAPTER 5

THE SECRET CODE OF CREATION

THREE-LETTER WORD

So what exactly is DNA? Well, the letters stand for 'deoxy-ribonucleic acid', a very special type of molecule. The name makes it sound much more complicated than it really is. The 'ribo' part comes from ribose, a type of sugar, because it's one of DNA's main building blocks. The 'deoxy' part means, as you might expect, that each ribose sugar has had an oxygen atom removed. And the 'nucleic' part comes about because it is found in the nucleus of a cell. So to sum up: DNA is a long molecule found in the nucleus of a cell, made of oxygen-lite ribose sugars.

Just in case all this is feeling a little bit unfamiliar, let's quickly remind ourselves what a cell is. You probably remember what one looks like from your school biology lessons; in case it helps jog your memory, I've drawn one opposite. If you scrape some cheek cells from the wall of your mouth, squirt a bit of dye on them and look at them under a microscope, you will see all the essential features of what biologists call eukaryotic cells.[1] There's a central nucleus, a sort of bubbly-looking jelly that surrounds it, and a cell wall around that. Your DNA, neatly packaged into twenty-three pairs of chromosomes, is in the nucleus. The jelly is what's called the cytoplasm – from the Greek for 'cell substance' – and the bubbles are all sorts of tiny gizmos that assist in the biological admin of the cell.

OK, so back to DNA. So what makes it so special? Well, first, it's very stable. As we shall see in a moment, it has a highly unusual structure that means it doesn't easily fall apart; we have recovered from hominin remains DNA that is tens of thousands of years old. Second, DNA is capable of holding information – in fact, it holds all the information necessary to clone you. And third, it's capable of making copies of itself. Without DNA, children wouldn't take after their parents, and there would be no way for evolution to work.

So to sum up: DNA is a very special type of molecule that carries information, a kind of blueprint for any given individual. It lives in the nucleus of the cell, where it is chopped up into smaller packages, called chromosomes.[2] Chromosomes come in pairs, and humans have forty-six in total. So how does

1 From the Greek, 'good kernel' (*eu* = good, *karyon* = kernel).

2 Different species have different numbers of chromosomes, and it seems to be fairly random as to how many they end up with. You might think that the more 'advanced' species would have more DNA and therefore more chromosomes, but one of the puzzles of biology is that it doesn't really work like that. Humans, for example, have forty-six chromosomes; a pineapple has fifty.

A Cheek Cell, A Chromosome and a strand of DNA

Cell wall

Cytoplasm

A Cheek cell

Nucleus

A Chromosome

A strand of DNA

it work? How do you get from the information in DNA to a walking, talking human body?

It's time to talk about protein.

STAYING ALIVE

Ask most people about protein and they'll probably mutter something about diets. They may even mention the famous Atkins Diet, beloved of noughties celebrities, where the emphasis is on foods high in protein and low in carbohydrate, and your breath smells like you've been drinking furniture lacquer.[3] More usually, though, they'll tell you that protein is one of the things we all need to eat, along with carbohydrate, fats, fibre, vitamins, minerals and Cheerios, and they'd be right.

Plants and animals, you see, are all about protein. You may already know, for example, that your muscles, bones, hair and skin are made of protein. But what you may not know is that it's proteins that get stuff done on a cellular level, too: from the enzymes that digest your food to the hormones that regulate your blood sugar to the haemoglobin that carries oxygen in your blood. It's a little unromantic, but it's not too much of an exaggeration to say that plants and animals are just bags of protein filled with water.

---

3 More on diets in Chapter 6; suffice to say that when you lose weight with the Atkins Diet, it's not because of any magical property of protein, or any harmful property of carbohydrate. It's because you consume fewer calories. In fact, the Atkins Diet puts your body into a state known as Ketosis, where you start burning off fat cells – hence the smelly breath.

Now proteins, like DNA, are long carbon-based molecules. But whereas DNA has ribose as its basic building block, proteins are made of chains of smaller molecules called amino acids. The jobs that a protein can do depend on its size, the exact combination of amino acids and – crucially – its shape. It is currently estimated that there are some 20–25,000 different proteins in the human body, but they are made up of just twenty different amino acids. Our bodies are able to make the vast majority of these amino acids, but not all. The ones our bodies can't make, we get from food. In other words, the reason we eat meat, eggs, fish and plant proteins is so that we can break them down into their individual amino acids, and then use those amino acids to build the proteins our bodies need.

And here's the best bit. For every protein in your body, there is a region of a chromosome that holds the codes for all the amino acids that are needed to build it, in the order that they need to be linked together. And the name for such a region might be familiar to you: it is called a gene.

MAPPING THE GENOME

I once met Francis Crick, one of the two men who famously figured out the structure and function of DNA. I was a graduate student at the Cavendish, and Rachel Weisz's mother Edith was giving a garden party for him at her home in Cambridge. Edith had known him since the sixties, when apparently he had been something of a butterfly on the Cambridge social scene.

It was in the summer of 1990, and he cut quite a dash, a silver fox with a pinch of Harry Potter wizard. I joined a conga-line of fellow students, all eager to unlock the secrets of the Universe, but kind of fluffed it by telling him all about how I had hallucinated by smoking banana skins when I was a teenager. This wasn't as random as it sounds, as I knew he was researching human consciousness, but all I can remember is him smiling kindly while no doubt wishing for a rather higher human consciousness to interact with.

Just as when I was introduced to Paul McCartney at a Steve Coogan gig and I shut myself in the loo rather than ask him what it was like to be in the Beatles, I probably did the right thing. Paul McCartney probably isn't the best person to tell you about the Beatles, just as Neil Armstrong probably isn't the best man to tell you about what it was like to walk on the Moon. Some people, after all, are just too close to the action to be able to appreciate quite how extraordinary their achievements are. And in the search to understand the origins of life, Crick and Watson were about as close to the action as it is possible to get.

Back in the early 1950s, everyone's hunch had been – correctly – that it was the chromosomes of a cell that were the seat of heritability. Chromosomes are made of DNA, and it was Crick and Watson, who had been shown a sneak preview of an X-ray diffraction pattern of a DNA molecule obtained by rival Rosalind Franklin, who managed to deduce that it had a double-helix structure, a bit like a twisted ladder. It's easier to draw than to describe so I've had a go above, next to my drawing of a cell.

Not only did Crick and Watson solve the riddle of DNA structure, they also cracked the DNA code. If DNA is like a ladder, then the 'sides' are made of alternating sugar and phosphate groups. Attached to each sugar is one of four different bases: adenine, thymine, guanine and cytosine. Let's not worry too much about what they are, because to understand the code we can just call them by their first letters, A, T, G and C. The bases join together to make the 'rungs'; A always joins to T and G to C.

And here's the truly astonishing bit. The bases form a code, and our bodies are capable of unzipping a DNA molecule right up the middle, reading that code and using it to make proteins. And, what's more, the code is unbelievably simple. Within a gene, the bases are grouped into words of three letters each, and each one of those three-letter words corresponds to one of the twenty amino acids our bodies are built from. In other words, a gene is basically a recipe for a specific protein; it simply tells the body which amino acids to put together and in what order.

You may have heard of the Human Genome Project, initiated by the same James Watson who discovered DNA along with Francis Crick back in the early 1950s. In simple terms, this landmark project consisted of determining the entire sequence of bases on a complete set of human chromosomes. The first complete sequence was obtained in 2003; ever since we have been working out exactly where on which chromosome each gene is and which proteins it produces.

All very well, you might say. Humans have DNA, packaged into chromosomes, and those chromosomes have genes that hold the blueprint for the proteins that make up our bodies. But here's the bit that's directly relevant to human history.

Everyone's DNA bears the hallmark of their ancestors and, with a bit of canny detective work, we can work back, not only through our immediate family, but thousands, millions and even billions of years into the past.

MITOCHONDRIAL EVE AND Y CHROMOSOMAL ADAM

If you compare the sequence of bases between two humans, there is very little variation. Asian, African, European, Inuit: we are practically clones of one another. There are about 3 billion bases in the entire human genome, most of them identical, but every few hundred bases a variable site crops up. Let's say for argument's sake that the beginning of Mike Tyson's first chromosome has the sequence:

CAGTTAGCTACTC

And Frank Bruno's has the sequence:

CAGTTACCTACTC

Then we can see that the seventh base along is a variable site, and Mike has a G where Frank has a C. Now let's say that, a few hundred bases later, another variable site occurs, with Mike's sequence reading:

TAGCTCTATCTAG

While Frank's reads:

TAGCTCCATCTAG

Again, we have found another variable site, and Mike has a T where Frank again has a C. You can see how a list of the variable bases can start to look like a signature for each individual, with Mike's variable bases beginning with the letters 'GT ...' and Frank's beginning with 'CC ...' Well, this type of signature is called a *haplotype*, from the Greek for 'single type', and it speaks volumes when it comes to the question of human origin.

Your haplotypes form a unique signature, made up of the exact bases you have at a given set of variable sites on the human genome. From one generation to the next, however, haplotypes are shuffled, with part of each chromosome in the child coming from the mother and part from the father. This mixing process, called *recombination*, is one of the benefits of sexual reproduction; it provides a wellspring of genetic diversity that natural selection can act upon and so evolve the species. Nevertheless, it's a bit of a drag when you want to trace your ancestry; after all, given enough generations, your haplotypes will be so shuffled that you'll hardly be able to tell one ancestor from the next.

But there are two areas of the genome where the shuffling we know as recombination doesn't occur, and they can give important clues, not only to an individual's genetic background, but also to the origins of our species as a whole. It's time to meet the Y chromosome and the mitochondrial DNA.

MIGHTY MITOCHONDRIA

A typical animal or plant cell is made up of a nucleus, packed full of DNA in the form of chromosomes, bobbing around in the jelly-like cytoplasm, surrounded by a cell wall. Remember I said the cytoplasm is packed with all sorts of handy gizmos? Well, one of those gizmos is a thing called a *mitochondrion* (Greek again, meaning 'thready granule'), which supplies the energy for the cell.[4] Mitochondria have their own DNA and – here's the important bit – are present in the female egg, but not in the male sperm. You therefore inherit your mitochondrial DNA solely from your mother, and it doesn't undergo recombination.[5]

THE CHIMPANZEE CLOCK

So how does mitochondrial DNA help us trace our ancestry? Well, DNA is good at replicating itself, but not perfect. Every time DNA gets copied, as it does every time new cells are made, mistakes creep in. Due to this process, even the haplotypes in mitochondrial DNA will mutate over time. The rate of change isn't nearly as fast as it is for chromosomal DNA, where recombination mixes the DNA of father and mother

4 Mitochondria closely resemble bacteria, and it's thought that they evolved from the bacteria that colonised the very first eukaryotic cells.

5 OK, to be really pedantic about this, sperm cells do have mitochondria, but only in their tails. Once the sperm has fertilised the egg, the tail drops off.

with each successive generation, but over roughly five thousand generations of human history, this mutation has a very noticeable effect.

By looking at an individual's haplotype, it's easy enough to place them within a wider collection of similar haplotypes called a *haplogroup*. The whole of mankind is split into around twenty mitochondrial haplogroups, all of which can be placed in chronological order. And, as you'd expect for a species that evolved in Africa and then spread out across the globe, there's one African haplogroup that appears to be the ancestor of all the other haplogroups on the planet.[6]

Putting an exact date on when each haplogroup emerged is currently an area of great interest, but the general principle is fairly easy to grasp. We know that we shared a common ancestor with chimpanzees, and we can work back to figure out what the mitochondrial DNA of that ancestor would have looked like. From fossils we know roughly when the human/chimpanzee split occurred; the latest evidence places it around 6 million years ago. So, knowing what our mitochondrial DNA looked like 6 million years ago, and knowing what it looks like now, we can take any intermediate DNA, analyse it and have a pretty good stab at calculating exactly when it first occurred.

This basic method is being refined all the time, as the rate at which mitochondrial DNA changes from generation to generation can be distorted by other processes such as natural

6 Haplogroups have been labelled A, B, C, etc., but rather unhelpfully they were named in the order that they were discovered, rather than chronologically. The African haplogroups are labelled L0–6. It's haplogroup L3 that appears to be the common ancestor of all the haplogroups outside Africa.

selection and population bottlenecks.[7] The picture I painted earlier of human migrations, beginning in Africa, running the coastal route to Indonesia, then heading back into Europe and across to the New World, is subject to constant tweaking and redating, but there is consensus on the broad outline and ever-increasing agreement on the fine print. And on one fascinating point there are no dissenting voices: we are all descended from a woman in Africa who lived around 200,000 years ago. She is known as the Most Recent Common Ancestor (MRCA), but I prefer her more poetic title: Mitochondrial Eve.

BACK TO THE GARDEN

So now you know: your mitochondrial DNA is inherited all the way down your mother's line and shows that you come from Africa. Not only that but, whether you are related to William the Conqueror or William Shatner, if you trace your mitochondrial family tree back far enough you will come to one woman, Mitochondrial Eve, who lived some 200,000 years ago.

To be strictly accurate, of course, Mitochondrial Eve wasn't

---

7 Natural selection removes harmful mutations, particularly with something like a mitochondrion, which either does or doesn't work. This means that there is less variation in present-day humans than there would be if harmful mutations survived, and so the real rate of change is greater than it appears. Population size has an effect because of random sampling. Common sense tells you that the smaller the random sample you take of a large population, the less likely it is to be representative of the whole. The relevance to the human story is that there may have been points in our evolution when our numbers got very low, distorting the rate at which our mitochondrial DNA has been changing. If that's the case, we need to account for that when we work out the age of Mitochondrial Eve.

the first woman who ever lived, just the luckiest of the first. It's likely that there were women who lived before her, and others who were her contemporaries, but their mitochondrial DNA happens to have died out while hers has survived. For that we have nothing to blame but chance; take any inheritable trait, and if a small population survives through enough generations, pure luck will eventually favour one expression of that trait over all the others. The first humans were by definition a small population, and it just so happens that only one strain of early mitochondrial DNA has made it down the millennia.

But the DNA in your mitochondria, inherited only from your mother, isn't your only link to the very first humans. There's another part of the human genome that you can trace back through the paternal line, all the way to one very lucky man, and that's the DNA in the Y chromosome.

BATTLE OF THE SEXES

Of the twenty-three pairs of chromosomes we all possess, twenty-two pairs are pretty much identical. The sex-determining chromosomes, however – known as the X and the Y – are a little bit different. As you probably know, if you are female you have the combination XX; if you are male you have the combination XY. When a female makes an egg cell, it gets one or the other of her X chromosomes. When a man makes a sperm cell, it gets either his X or his Y. In other words, the sex of a baby is effectively determined by the sex of the sperm that fertilises the egg. No doubt, at some point, some clinic will figure out a way to separate male from female sperm and be able to control the sex of human babies, but for the moment – whatever anyone tells you about boys running in their family – the process is exactly fifty-fifty.[8]

As I mentioned earlier, the twenty-two 'normal', non-sex-determining chromosomes get shuffled during sexual reproduction in a process called recombination. This has the effect of mixing the traits of the parents in the child, which is why you might easily end up having the shape of your mother's eyes but your father's eye colour. Each successive generation becomes a sort of chromosomal melting pot, with all traits switched on at some point or other and therefore subject to natural selection. Advantageous traits will be favoured and therefore spread; similarly, harmful traits will fall by the wayside.

It's a similar story for the two X chromosomes in a newly fertilised female egg. They will also have undergone recombination, meaning that they are a mix of the X inherited from the mother's egg and the X inherited from the father's sperm. Again, this is a very healthy thing, meaning that any dodgy bits of DNA on an X chromosome will eventually get expressed and weeded out by natural selection. But the Y chromosome, as we shall see, is a horse of a very different colour.

Just a glance at a typical Y chromosome will tell you something's up. The X chromosome is the eighth largest, but the Y is lean and mean. It has a compact 60 million base pairs and

---

8 As you may guess, the 'boys run in my family' thing is yet another example of random sampling. The greater the number of children, the greater the chance that half will be girls and half will be boys.

contains only 86 genes, whereas the X chromosome has a strapping 153 million base pairs and approaching 2,000 genes. What's more, they really don't get along; X and Y chromosomes are so mismatched that they don't get mixed during sexual reproduction.

Now it's easy enough to see a good reason why the Y chromosome might be shorter than the X; after all, women don't carry that chromosome at all, so you don't want any truly crucial genes on it other than those that are responsible for male traits. And it's also easy to see why you wouldn't want the X and the Y to recombine, as you'd get a mixing of male and female traits on two new chromosomes that were, as a result, neither Arthur nor Martha. But it does mean that harmful mutations on a Y don't get expressed as often as they do on an X, and therefore aren't so easily removed by natural selection. It also means the feistiest, fittest Y chromosome on the planet can still go out of business through no fault of its own. After all, no matter how square your jaw or how firm your bicep, chance may rear its head and ensure you sire no male offspring, with the result that all your male genes instantly bite the dust. In every generation, the genes of a whole host of virile men, who happen by luck of the draw to have fathered squadrons of girls but no boys, are lost to us for ever.

In other words, the battle of the sexes is very real, and at the chromosomal level you'd have to put your money on the girls. But though the lack of mixing between the male and the female chromosomes brings a unique set of challenges for the Y, when it comes to tracing human ancestry its standoffishness is a positive boon. Because just as mitochondrial DNA is handed down from mother to daughter, the Y chromosome is a direct link to the first male humans to walk the planet. And just as mitochondrial DNA leads back to one woman, so Y chromosomal DNA leads back to one man: Y chromosomal Adam, who lived in Africa roughly 140,000 years ago.

LONG-DISTANCE RELATIONSHIP

Just as with mitochondrial DNA, Y chromosomal DNA can be grouped into haplotypes, each of which bears a unique signature of mutations. And, interestingly, those haplotypes tell a very similar story: of a group that left Africa and populated first the Old, then the New World. And just as the haplotypes in your own mitochondrial DNA will have arisen somewhere along that journey, be it in Africa, Asia or the Arctic, so will the haplotypes in your father's Y chromosome. Imagine that; you can do all the research in public records that you like, but the order of bases on your chromosomes can tell you the deeper history of where your ancestors came from.

Of course, there's the small matter that Eve appears to have been the greatest cradle-snatcher in the history of mankind; she was, after all, some 60,000 years Adam's elder. Needless to say, they weren't a couple. The most recent common ancestor of all mitochondrial DNA did not have to meet the most recent common ancestor of all Y chromosomal DNA in order for the human race to prosper. Eve would have had a male partner, but sadly his Y chromosome, along with all the Y chromosomes of his kin, has died out.

In other words, Y chromosomal Adam is one of many fathers, just as Eve is one of many mothers. What we are talking about when we say we can trace our ancestry back to him is just the lineage of our Y chromosomes. The DNA in a man's Y chromosome is just a fraction of his total DNA – much less than a fiftieth, in fact – as is the mitochondrial DNA all modern men and women inherited from Eve. All the rest of your DNA originated with one of the early humans, too; it's just much harder to work out whose it might have been because it has been so scrambled. In fact, every gene on every chromosome in your body must have its own most recent common ancestor. The time may yet come when we can trace the family tree of our elbows, upper left incisors and belly buttons.

Because when it comes to DNA, the people involved are only part of the story. We are about to look at evolution from your DNA's point of view, and I'm afraid it may make for some chilling reading. Because as far as your genes are concerned, human beings are just handy lifeboats. For a gene, the game is simply to make as many copies of itself as possible, and for that all it needs is a host who mates and produces viable offspring. Let's go through the looking-glass into the mirror world of the gene.

THE SELFLESS GENE

To see what I mean, let's just remind ourselves what we're dealing with. Human DNA, you will remember, is packaged up into twenty-three pairs of chromosomes. Each pair of chromosomes is pretty much identical, apart from in men, when one pair will consist of the mismatched X and Y sex chromosomes. During sexual reproduction, the chromosomes of both parents are mixed to produce a new set of twenty-three pairs, unless they produce a boy, in which case the father's Y chromosome and the mother's X chromosome will be passed on pretty much intact. All this we know.

You will also remember that each chromosome is a very long helical molecule with two intertwined carbon-based backbones. Connecting the two backbones are the four 'bases' – A, T, G and C. The incredible 'secret' of DNA is that these four letters form a code, with each three-letter word corresponding to an amino acid. DNA is therefore very much like a recipe book for making proteins; it records which amino acids to combine and in what order.

Well, almost, because there's a subtlety here. The 'coding' bits of DNA – the bases that will eventually be transcribed into amino acids – are scrambled up within the gene. The bits with the code are called *exons* and the bits in between are called *introns*. Part of the process of manufacturing a protein for a given gene involves snipping out all the introns and stitching together all the exons. To see what I mean, imagine a few words of text inserted among gobbledegook:

thegdtdyquickjdudybrownksysghsfoxjssustjumpsjstsgtsover-jsjsthehslazyksdog

In this analogy, 'gdtdy' is an intron, and 'the' and 'quick' are exons. Part of the glory of having introns is that you can splice

THE SECRET CODE OF CREATION

the exons together in different ways, and therefore make more than one protein from a single gene. Just as you can snip out 'gdtdyquickjdudy' from our text – along with the other introns – and splice what's left into a sentence that still makes sense but has a slightly different meaning:

thebrownfoxjumpsoverthelazydog

In other words, you can use the same gene to produce several different proteins, maybe as many as ten for a given gene.

By and large we humans are clones. The 3 billion or so bases in my DNA are pretty much identical to yours, and yours are pretty much identical to those of every member of the Beatles, even Ringo. That said, every now and then a variable site does crop up. In other words, instead of a G at a particular site, maybe you get an A, or maybe instead of a C you get a T. Of the 3 billion bases in the entire genome, only around 10 million are variable, meaning that on average a variable site crops up only every three hundred bases or so. You will also remember that when we make a list of these variable bases, we make a kind of signature for that individual's DNA, called a haplotype.[9]

Earlier we marvelled at how haplotypes in mitochondrial DNA and Y chromosomal DNA can give us clues as to our ancestry, since these parts of the genome don't undergo

9 By the way, variable sites (also known as SNPs or Single Nucleotide Polymorphisms) are important but they aren't the only source of genetic variation. From one generation to the next you also get copying errors, and bits of DNA can be switched on and off by mistake, all of which results in variation between individuals.

recombination – in other words, mixing – during sexual reproduction. By sampling haplotypes from several different races in different parts of the globe, we can put them into family groups, called haplogroups, and even work back to calculate what some of the earliest human DNA looked like. But there's also work to be done with the haplotypes of the other chromosomes. After all, if you think about it, if you see the same sequence of variable bases cropping up in more than one person's chromosomes, despite all the mixing going on during sexual reproduction, it is like a big flashing neon sign that says, 'This bit of DNA is doing something important.'

To see what I mean, imagine a haplotype as a deck of cards. Recombination means that in each successive generation, you cut the pack. A large number of generations later, we'd expect every pack to have been cut so many times that the cards become randomly mixed. How surprised would you then be to examine each deck and see that in every case the four aces were all side by side?

If you were a card player you would deduce foul play, and if you were a biologist, and seeing something similar in haplotypes, you would deduce natural selection. The chances are that if a large number of people all share the same pattern of variable bases, those variable bases represent an increase in fitness for whoever carries them. But how could changing the odd T base, say, to a C base produce any sort of evolutionary advantage?

You may have guessed that I'm building up to something, and you'd be right. We now have all the pieces of the puzzle that we need to see the last 10,000 years of human civilisation from the point of view of a single mutant gene. You are about to hear a fascinating history that intertwines the domestication of cattle with the success of Indo-European languages: nothing less than a genetic mutation that has experienced a surge of popularity due to human culture. If you ever doubted that human evolution is going on all around you, you need to hear the story of a variable base in one of the introns of gene MCM6 on Chromosome 2.

MILKMEN OF THE STONE AGE

The chances are that if you are a northern European, or a North American of northern European extraction, you drank milk this morning and didn't give it too much thought. If you are African or Chinese, the chances are that you didn't; in fact, you probably drink very little fresh milk at all. And if you don't drink milk you are in the majority, because something like 65–75 per cent of the world's population are unable to digest milk in adulthood. They are what scientists call lactose intolerant.

Lactose, you see, is the complex sugar in milk, and it needs to be broken down into two smaller sugars, galactose and glucose, in order to be digested by the body. The body produces an enzyme called lactase that helps us do this. All babies have this enzyme; humans, after all, are mammals, and mammals produce milk to feed their young. Back in the bad old hunter–gatherer days of the human race, long before farming kicked in around 10,000 years ago,[10] human babies were weaned from their mother's milk like any other mammal species and when they were adults ne'er a drop of the stuff would pass their lips. Everybody was lactose intolerant.

Now there's a certain amount of confusion surrounding this topic in non-scientific circles, at least there is in the non-scientific circles I move in. Many northern Europeans seem to believe that lactose intolerance is one of those 'fad' allergies that has recently emerged in the West. To them, the ability to drink milk as adults may seem normal, but if you really stop to think about it, it is truly odd, almost vampiric. Because – bar one or two *Little Britain*-style exceptions – grown men and women don't drink their mother's breast milk, which is of course what the lactase in infant bodies is designed to digest. Adults drink the breast milk of dairy cows. That's right. The milk that dairy cows produce to feed their own cute little baby cows is being harvested and drunk by great, big, hairy, ape-descended *Homo sapiens* men and women. If you chanced upon a crowd of woodland squirrels milking a hapless badger you'd be truly disgusted, but that, my friends, is very much the kind of game we're into. And somehow we have convinced ourselves it's normal.

10 It is the beginning of agriculture that defines the Neolithic, or 'New Stone Age'. It began at a different time in different parts of the world; the very first farmers are thought to have domesticated grains independently in China and the Levant around 10,000 years ago. In northern Europe the Neolithic started about 7,000 years ago and ended 4,000 years ago. It was followed by the Bronze Age (roughly 5,000–3,000 years ago in northern Europe) and then the Iron Age (roughly 3000–1500 years ago in northern Europe). After that, of course, came the Middle Ages and finally the Modern Age.

IT'S IN THE GENES

OK, so lactase is the enzyme in *Homo sapiens* – and in mammals, as well as in a great many other species – that breaks down milk sugar, aka lactose, into smaller sugars that can be digested by the body. An enzyme is a type of protein that aids digestion and, as we know, all the body's proteins are coded in DNA. And it so happens that we have found the gene that codes for lactase: it is called LCT, and in humans it sits on Chromosome 2.

Human genes, by the way, are named rather soberly after their function, hence LCT after lactase. Genes found in other species, such as *E. coli*, fruit flies and mice (all common subjects of genetic research), have tended to have slightly more fanciful names. So, for example, we have the fruit fly gene 'Cheap Date', which increases sensitivity to alcohol, and the 'Ken and Barbie', which internalises genitalia. My personal favourite is 'INDY', which increases fruit fly lifespan, referring to the diseased serf in *Monty Python and the Holy Grail* who doesn't want to be put on Eric Idle's cart, protesting, 'I'm not dead yet.' Back in the early days of genetics, of course, most of the work was being done on non-human species; the odd madcap gene name kept things interesting in the lab and harmed no one. But because all species are related, and DNA is also transmitted across cell lines by viruses and bacteria, genes that are common in, say, fruit flies are often found doing something similar in human beings. Nobody wants to hear that they have the 'Cheap Date' gene, so there has been a rearguard action by human geneticists to get the fruit fly lot to fall into line and start calling their newly discovered genes something sensible. Which must be a bit galling when you've been staring at fruit flies for six hours and you need a bit of light relief.

Anyway, back to the plot. In their search for a mutation of the LCT gene to explain the existence of those among us in northern Europe who are able to drink milk in adulthood, scientists found nothing. They widened the search to include other genes near LCT and found something very interesting indeed. In gene MCM6 there was a single variable base, LCT-13910 C/T, which was strongly linked to lactose tolerance in adulthood.[11]

The exons, or coding regions, in gene MCM6 hold the recipe for a ring-shaped protein called minichromosome maintenance complex 6, which plays an important role in DNA replication. That's by the by, though, because the single variable base that is connected to an ability to digest milk sits inside one of the introns on this gene, about 14,000 bases away from the LCT gene. Researchers found that if members of their northern European sample group had a T base at this site, they were able to digest milk in adulthood. If they had a C at the same site, they probably had toast for breakfast instead.

A link, of course, isn't the same thing as a cause. Every time I have a cup of coffee I have a spoonful of sugar in it, but it's not the sugar that gets me buzzing like a handsaw, it's the

---

11 Genetics is a frontier town, and there is a certain amount of lawlessness when it comes to describing the location of a specific base. The notation I've used is one of the more common ones, and basically says the base we are talking about is a T for thymine and it's 13,910 bases along from the LCT gene.

caffeine. Likewise, the fact that we have found a variable base that is linked to an ability to produce lactase in adulthood doesn't mean that the one is definitely the cause of the other. The human body is extremely complex, and exactly how changing a single C base to a T base might alter your ability to digest milk is a tough nut to crack. It is easy enough to understand the effect of a single variable base in the coding DNA of a gene; it would change an amino acid and therefore produce a slightly different protein. Our understanding of how the DNA in introns might control genes, however, is less complete. And in any case, the variable base that may be switching on the LCT gene isn't even in an intron of the LCT gene itself; it's in an intron of another gene entirely.

Nevertheless, setting aside those words of caution, there is some very interesting circumstantial evidence to support the idea that an ability to digest milk is something that evolved in northern Europe after the invention of dairy farming. For a start, it turns out that the areas with the highest levels of lactose tolerance match the location of Neolithic farming sites. Second, we have tested the DNA of Neolithic skeletons from northern Europe, which, at around 7,000 years old, pre-date dairy farming, and none of them has the LCT-13910 T mutation, which suggests that dairy farming came first and the mutation followed in its path. And third, the DNA of late Bronze Age skeletons, dating from around 3,000 years ago, shows that around half of them do have the mutation. All of which supports, but doesn't prove, the idea that dairy farming encouraged the spread of the LCT-13910 T mutation in northern Europe.

Oh, and as Columbo would say, one last thing. Remember

what we said earlier about haplotypes? When a large number of people have the same sequence of variable bases, known as a haplotype, it's a good indication that somewhere in the underlying DNA is something that improves fitness. And because our chromosomes tend to get shuffled with each new generation, the longer the shared haplotype, the shorter the time the beneficial mutation that it contains must have been around. Northern Europeans with the LCT-13910 T mutation have been found to share one of the longest haplotypes known in human DNA, again tying in with the idea that it became widespread within the last few thousand years, after the introduction of dairy farming.[12]

THE LANGUAGE OF THE GENOME

So here's where we've got to. If our hypothesis is correct, a single variable base in a non-coding region near the lactase gene helped create a new breed of northern Europeans who were capable of digesting cow's milk. As dairy farming spread, so did these milk-drinking vampire apes, until the present day when something like 95 per cent of northern European men and women carry the mutation as well as a large chunk of the DNA around it.

12 The shared haplotype for northern Europeans with the LCT-13910 T mutation on Chromosome 2 is around 1,000 variable bases, corresponding to an underlying chunk of DNA that is about 2 million bases long. That's just under 1 per cent of the chromosome: evidence of strong recent selection.

I love this story for so many reasons that it's hard to know where to start. To begin with, it's all too easy to see evolution as something that happens to other species, but in this case it's clearly very close to home. Not only that, but we tend to think of natural selection as requiring endless millennia of more or less unnoticeable change, but here we have a mutation that has spread to unbelievably high frequencies within a few thousand years. And, as the capper, the driving force that turned so many of us into milk-drinking mutants was a change in culture in the form of dairy farming. In what other ways has our society shaped our evolution, and in what ways is it shaping it now? Are those of us who can't program in HTML slowly being outbred by those who can? And is our evolution being speeded up by our increasing population size, which brings with it a greater chance of advantageous mutations? It's hard to stop your head from spinning.

Another reason is that it provides a fascinating glimpse of evolution seen from a gene's point of view. As we know, the name of the game for an individual gene is to provide its host with a means of having more offspring. And in Neolithic culture, when calories were at a premium, a mutation that gave its carrier a renewable bovine energy source would have been a very big advantage indeed. Genes have their own battle for survival, and in this case LCT-13910 T appears to be winning. A sizeable clump of DNA associated with the trait of adult milk-drinking is rooting out all opposition, much as the aggressive grey squirrel has replaced the placid red. And isn't something similar going on with every trait in the human body, be it eye colour, whiteness of tooth, or proficiency in pub darts? Our bodies are a battleground, and any gene that vastly improves our chances of having offspring will turn the field. As far as our genes are concerned, you and I are simply a means to an end.

I also love the flights of fancy this kind of science can lead you on. After all, if a milk-digesting mutation was beneficial enough to sweep northern Europe in a few millennia, what other advantages did it bring? Could it be, as some people think, that copious calories from highly mobile cattle helped carriers of this mutation to spread while non-carriers were tied to their grain harvest? And, specifically, was this the advantage of the Kurgans, the inhabitants of the grasslands between the Black Sea and the Caspian Sea, who many linguists believe are the progenitors of the Indo-European group of languages? Can it be possible that the languages of Europe, Russia and Asia spread because some freak cattle herdsman on the Pontic–Caspian steppe developed a mutation whereby he could digest milk?

And finally I love this story because it presents the perfect snapshot of where we are with genetics, and a clear signpost to where we might be heading. In some ways the coding regions of DNA were the easiest place to start in understanding the human genome. After all, they are written in a language that we can understand and they produce a tangible product, a protein, which we can isolate and investigate. But it's the non-coding regions – the large gaps between genes and the introns that lie within them – that hold the greatest mystery. After all, a mouse and a man have roughly the same genes, so the reason most mice don't hold down jobs or shop online has to be something to do with the way those genes are controlled.

Back when I first studied evolution, in the 1980s, it was hard to imagine how a single random mutation in one of the bases of the DNA of an organism could produce a large enough change in its fitness for it to be selected by an environment in which there were all sorts of other random life-threatening events like earthquakes, droughts and disease. The astronomer Fred Hoyle put this eloquently as a 'signal' and 'noise' problem: how could a small 'signal' in the form of a selection advantage be heard above the 'noise' of everyday existence? Yet in the case of mature specimens of *Homo sapiens* drinking the milk of *Bos taurus* we are talking about the change of a single base, not in the coding DNA of the relevant gene, but in the non-coding DNA that controls it. LCT-13910 T may be only part of the picture as regards lactose tolerance, but it is a tantalising glimpse of how small mutations in the controlling regions of our genome can produce big changes in our fitness. That's a huge slap on the back for the theory of evolution.

Mapping the human genome was a landmark in genetics, but the task ahead of us is enormous. The more individuals we map, the more we discover that the genome is far from fixed; it is constantly shifting, changing from generation to generation and quite possibly throughout our lifetimes too. Variation may come from many sources: viruses that write themselves into our DNA, copying errors or mutations of single bases, to name just a few. Yet as complex and plastic as our genome is, progress is constantly being made, often at a simply breathtaking pace. The prize, as we all know, is not just the understanding of diseases – so many of which, like cancer, cardiovascular and autoimmune disease, diabetes and psychiatric illness, have a hereditary component – but the chance to alter the genome itself and launch a new wave of advantageous human mutation. If human evolution has speeded up with the progress of civilisation, as it certainly seems to, imagine how much quicker it will gather pace when we can tinker about with the genome itself.

And then, of course, there's another world beyond that of DNA, which I have left pretty much untouched in this book, but which promises even deeper understanding of the molecular basis of life: that of RNA. A recipe book is useless without a chef, and RNA is the middleman between DNA and protein, but, unlike DNA, which stays locked inside the cell's nucleus, RNA can travel in and out of the body's cells at will.[13] In fact, to call it a chef isn't really doing it any sort of justice; it's chef, kitchen, kitchen staff, restaurant and waiters all rolled into one. Essentially RNA reads the code on the DNA molecule, makes a copy and travels out of the cell into the body where it assembles the required amino acids into the relevant protein. These proteins then do all the things a body needs to do in order to thrive, such as forming tissues and organs, and breaking down food into fats, carbohydrates and sugars.

One of the most fascinating challenges of biology is the search for the origins of life, and in RNA there is the real

---

13 For the budding chemists, RNA is a close relative, molecularly speaking, of DNA. Both are called 'nucleic acids' because they were first discovered in the nucleus of cells, and the basic building block of RNA is also ribose, hence its proper name, 'ribonucleic acid'. DNA is more or less the same thing, but missing a hydroxyl group, hence its formal title 'deoxyribonucleic acid'. Don't worry if you don't know what a hydroxyl group is; the point to grasp is that removing it makes DNA much more flexible than RNA, and therefore capable of binding with a second strand into the famous – and ultimately more stable – double helix.

possibility that we are close to the source. Some current research hints at RNA being an older molecule than DNA: a prototype, self-replicating, single-stranded molecule that lacked the chemical stability that came later with double-stranded DNA. Which leaves us with the possibility that our first ancestor was a rather unstable, overactive sugar molecule. Now stick that in your family album.

CHAPTER 6

LET THEM EAT CAKE

KITCHEN NIGHTMARES

In May 2008 I received a challenge. Would I be prepared to follow in the footsteps of esteemed celebrities such as DJ Chris Moyles and TV presenter James May, and have a cook-off with Michelin-starred chef Gordon Ramsay on his Channel 4 show the *F Word*? As if that wasn't daunting enough, the competition was to take place on Gordon's home turf, in the ultra-high-spec kitchen of the *F Word*'s 'glamorous and bustling restaurant', with our dishes then being blind-tasted and marked by independent judges.

Many men would have quailed and fled. And that's exactly what I did. After all, Gordon is as well known for his confrontational presenting style as he is for his perfectionism in the kitchen, and his programmes invariably involve him psychologically scarring some celebrity or other while they attempt to measure up to his unfeasibly high culinary standards. I am no fool; I told my people in no uncertain terms to tell his people that the whole thing was out of the question.

And yet several days later I found myself reconsidering. My love of food ranks right up there with my love of science, and Gordon Ramsay has long been one of my heroes. I had met him once before, back in the 1990s, when Alexander Armstrong and I had been regular performers on the Radio 4 programme *Loose Ends* and Gordon had appeared as a guest. His sparky answers to Ned Sherrin's somewhat prickly questioning had been the unexpected highlight of the show, and Alexander and I had needed no further excuse to try his fine handiwork at his first London restaurant, Aubergine. He had rapidly become one of my favourite chefs, as well as presenting one of my all-time favourite shows in the form of *Kitchen Nightmares*. The chance to meet him and experience the Ramsay brio at first hand was too tempting to miss.

INSIDE INFORMATION

I am lucky enough to have several friends who work in the restaurant business. Well, all right, it's probably not luck, it's greed and the fact that I spend ridiculously large amounts of time eating in restaurants. One of those friends is Henry Dimbleby, co-founder of the Leon chain. Henry, like me, also happens to be a trained physicist, and has devoted his professional life to understanding as much as possible about the art and the science of cooking. I knew he would understand my desire to trounce Ramsay and thereby gain his respect. Not only that, but Henry could have crucial insider knowledge of any soft spots in Gordon's formidable armoury.

'He can't bake' was Henry's verdict, following shady lines of enquiry that went right to the heart of the Ramsay empire. 'None of those big chefs can. They are great at that grandstand, show-stopper savoury stuff, but they think that baking is for girls.' Surely it couldn't be that simple? 'Think about it. They all train under the French system, so English cake is a foreign land. Plus most of them are psychologically complex and the one person in their household who baked was their mother. Why does any man become a chef other than to impress his mother? Pull it off and it will be like a missile fired right into the heart of the Ramsay Death Star.'

It was an audacious plan, but it just might work. Baking is, after all, chemistry that you happen to do in the kitchen. I may not be an experienced cook, but I am a trained experimental scientist. Cake-making could be just the thing to play to my strengths and Gordon's weaknesses. I called the *F Word*'s researcher. I would face Gordon in a Victoria sponge bake-off. As I hung up the phone, I was overcome with a wave of nausea. With barely an oven-heated ready-meal to my name, had I really just accepted a showdown with a world-class chef on national television? And yet, some still voice inside me was quietly confident. Maybe, just maybe, with the right planning and research, Gordon might find himself in a kitchen nightmare all of his own . . .

AMOROUS ATOMS

The fact that Victoria sponges exist at all, or that we exist to make and eat them, is really all down to the remarkable neediness of atoms. Unable to stand their own company, they are constantly getting into all kinds of relationships, suitable and unsuitable, with just about any other atom that happens to be close to hand. The name for these groups of atoms is, of course, molecules, from the Latin *moles* meaning small mass. And molecules can have all sorts of different properties above and beyond those of the atoms they are made up of.

For a start, molecules can be large, like proteins, or small, like carbon dioxide. Roughly speaking, once you know the size of a molecule and its temperature, you can take a stab as to whether it will be a solid, a liquid, or a gas. Here on the surface of the Earth, where the temperature tends to average around 15°C, small molecules tend to be gases. Classic examples would be the gases in the air: nitrogen, the most common, is made up of just two nitrogen atoms; oxygen, the next most abundant, is made up of a mere brace of oxygens. Water molecules, being made up of three atoms, are slightly bigger, and water, as you know, is liquid at most Earth temperatures.[1] In fact, one of cosmologists' chief criteria for a planet to have

1 Chemical formulae for molecules essentially work by telling you how many atoms of each type are present. Hence nitrogen gets the symbol N_2 and water goes by H_2O. What they don't really tell you is which atom is stuck to which; this is where all those convoluted chemical names come in, like '1,2-dimethylcyclopropane', which is a spirited attempt to provide a sort of road-map as to the shape of a molecule that has the chemical formula C_5H_{10}.

Earth-like life is that water should be liquid on its surface. The point being, I suppose, that if the average temperature on the surface of our planet was 1,400°C rather than 15°C, we would have long since turned into gases and wafted off into the ether.

There's no limit, really, to how big a molecule can get. A DNA molecule can have as many as 15 billion atoms, which is really going some. The bigger a molecule is, of course, the more likely it is to be a solid at room temperature; the smaller it is, the more likely it is to be a gas. Many of the foods we eat, such as carbohydrates and proteins, are very much solid at Earth temperature and, as we shall see, this poses unique problems not only for our digestion but also our gustation. After all, how am I supposed to extract the amino acids from a lump of raw egg and get them into my bloodstream, and why should I bother when said raw egg hardly tastes of anything? These, fellow chef, are just two of the problems that cookery sets out to try to solve . . .

ACTION AND REACTION

Not only are atoms always getting into relationships with other atoms, but they are always, always on the lookout in case something better comes along. Atoms A and B, say, may have just tied the knot in the form of molecule AB, but the mere sight of atom C will be enough for A to ditch B and form new molecule AC, leaving B completely in the lurch. This type of wanton promiscuity is known in the world of chemistry as a reaction. In fact, you could make an extremely broad generalisation and say that the whole of chemistry is really about trying to predict which new lasting relationships will emerge when a given bunch of molecules decide to party together.

As it is in human love affairs, so it is with atoms: the readiness with which partner-swapping goes on depends on how happy the participants are in their existing relationships. Sometimes atoms need little excuse to part company from their existing set-up because the new relationship promises so much more stability than the old one. Add sulphuric acid to copper oxide, for example, and the sulphate will immediately fall into the arms of the copper, producing that wonderful bright blue salt, copper sulphate. The oxygen from the oxide and the hydrogen from the acid, meanwhile, will be left to pick up the pieces and struggle on as water.

Sometimes, however, such swaps don't occur readily, but will happen with a little gentle persuasion. They may, for instance, take a bit of encouragement in the form of energy supplied as heat, or the presence of a matchmaker molecule who encourages partner-swapping without actually partaking themselves, also known as a catalyst. And when it comes to the series of chemical reactions known as digestion, it's the body's own catalysts, known as enzymes, that really make it all possible.

INDUSTRIOUS ENZYMES

When we eat, our bodies are faced with two major tasks: how to break down carbohydrates and fats into sugars and fatty acids that can be burned to provide us with energy, and how to break down proteins into amino acids that can be used to build new proteins. To be strictly accurate, at a pinch our bodies can burn amino acids to make energy too; in fact, if you fail to eat enough carbohydrate and fat to keep your motor running, your body will start to break down even your own muscle fibre into amino acids so that it can burn those instead. And when I use the word 'burn', I really do mean that the sugars and fatty acids from digested food undergo a chemical reaction where they are combined with oxygen to release energy in a way that has much in common with the burning of a log in the grate. That said, it will not have escaped your notice that we don't have to set light to our pies before we eat them. Thanks to the enzymes in our digestive tracts, the burning reactions that power our bodies can take place at much lower temperatures than they would do otherwise.[2]

As it happens, we've already met one of these enzymes in the last chapter. Remember lactase, which helps the body to break down a large, otherwise indigestible milk sugar called lactose?[3] Well, lactase is just one of an army of similar proteins, coded in our DNA, all of which are designed to get

2 You may be interested to know that this 'burning' happens in the mitochondria of the cells. And it is mitochondrial DNA, you will recall, that is inherited only from your mother and therefore provides a handy way to trace your matrilineal ancestry. You see: your hard work is paying off. It's all coming together.

3 As ever I am aiming for the broad strokes, so I've used the words 'sugar', 'starch' and 'carbohydrate' interchangeably in this chapter, but there are of course important differences that the martinet in me can't let pass without comment. The word 'sugar' is generally used for small, sweet-tasting carbohydrate molecules like glucose, so something bigger like lactose is really on the boundary between what you might call a sugar and what might otherwise be called a carbohydrate, especially as it's not sweet to the taste. Starch, strictly speaking, is a subdivision of carbohydrate, being the type produced by plants. Now you know that, feel free to forget it.

to work on breaking down fats, proteins and carbohydrates into smaller molecules that can enter the bloodstream and be transported throughout the body. Unlike other animals, of course, we humans have found a way to maximise the effectiveness of our enzymes through the extraordinary invention of cookery.

PIMP MY ENZYME

So why cook? Well, there's the obvious reason that it destroys harmful bacteria, but that's just a small part of the story. One of the main problems with many raw foods is that they are tough. Maybe that's fine for a gorilla with all day to nibble at bamboo shoots, but we humans simply don't have the time or the will. Our brains are energetically expensive items, and we need ways of getting more calories and nutrients for less effort. After all, if you just sit in your favourite armchair, doing nothing more than reading an informative and entertaining popular science book, in the process your brain will use up a whopping 25 per cent of the energy you get from your food. Cooking softens the things we eat, making them easier to chew and swallow, and there is evidence to suggest that the invention of cooking was a key factor in enabling increased brain size in early humans.

In fact, the leap seems to have occurred around 1.9 million years ago, with the appearance of *Homo erectus*. Earlier hominids had smaller brains: 600 cubic centimetres for *Homo habilis*, for example, compared to 900 cubic centimetres for

Homo erectus. It's telling that *Paranthropus robustus* – a dead-end branch of the human family tree from around that period that didn't make the cut in evolutionary terms – had very pronounced skeletal adaptations for grinding up raw plant foods, such as heavy jaws, huge teeth and crests that ran along the midline of their skulls where hefty chewing muscles could attach. In contrast, signed-up members our own genus, *Homo*, had smaller faces and more delicate jaws despite being larger in terms of overall body size. All of which lends some weight to the theory that *Homo* was dining on high-quality cooked foods while *Paranthropus* was busy chewing on uncooked roots.

And then there's the problem that there are a lot of raw proteins and carbohydrates that we don't have the enzymes for, so even if you do manage to chew them into small pieces, there's not much that your body can do with them anyway. Eat a raw egg, for example, and you'll digest only a portion of the protein; eat a raw potato and you'll metabolise only half of the starch.[4] Cooking, in other words, starts to break down key structures and bonds in the proteins and starches, turning them into a form that your digestive enzymes can get to work on. That means many more foods to choose from, and a huge evolutionary advantage when it comes to survival.

And finally, of course, there's taste and texture. Cooking can alter the physical structure of the food we eat, and give us all sorts of pleasure simply from the way that it behaves in our mouths:

---

4 When you cook an egg, you digest about 94 per cent of the protein, whereas if you eat it raw, only around 60 per cent is digested and the rest is lost. Similarly, you digest about 95 per cent of the starch in a cooked potato, but only around 51 per cent of that in a raw one.

crunching and crackling, for example, or creaming and coating. And, most importantly of all, as we shall soon see, if you raise foods to high enough temperatures, you can initiate a whole family of chemical reactions where proteins dance with sugars to create generation upon generation of delicious-tasting flavours.

A QUESTION OF TASTE

Have you ever thought about why some foods taste good and others don't? Why, for example, does a lick of ice cream taste sweet while a mouthful of rhubarb leaf tastes bitter? The answer, quite simply, is that our taste buds have evolved to steer us towards anything that is a good energy source (such as sugars) and away from anything that contains toxins (the one in rhubarb leaf is oxalic acid, a useful industrial cleaning agent. Not surprisingly, ingesting rhubarb leaf can make you very ill, so please don't.). For our ancestors, energy-giving foods were of premium importance and our palates evolved accordingly. In other words, don't blame the chocolate bar on your weak will; blame it on the scarcity of energy-giving sugars in the world of our ancestors.

As well as sweet and bitter, we currently recognise three other tastes: sour (think of acid-rich fresh fruit), umami (a response to an amino acid called glutamate, so think of a meaty broth) and salt (no prizes for guessing what to think of there). I say currently, because umami was officially added to the list only in the mid-noughties, and there is some new research to indicate that we can taste fat too. By the time something gets in our mouths, we basically just need to know whether to swallow it or spit it out, and our taste buds are there to tell us that it's OK to swallow sugar, fruit, amino acids and salt, but it's best to hack up anything that might be poisonous.[5] Yet, as we all know, foods come in a glorious variety of flavours – far more than five. Odd as it may seem, to really understand taste we need to look further than the taste buds, for, as we are about to see, the greater part of what we think of as taste is actually smell.

SIZE MATTERS

Our tongues may have between 2,000 and 8,000 taste buds with around one hundred taste receptor cells in each that get excited when they come across one of five different tastes; our noses, on the other hand, have something like 5 to 10 million olfactory cells capable of responding to hundreds of different odours. Our taste buds therefore give us a basic message about what it is that we are eating, together with a certain amount of pleasure if that thing happens to be a useful food group. It's our noses, on the other hand, that really capture the fine detail. For a food to be truly delicious, it needs to have fragrance.

---

5 Fresh fruit, of course, is a good source of water-soluble vitamins, fibre and water; amino acids are necessary for our DNA to encode new proteins; and salt is vital to our nerves, muscles and blood cells. You might also hazard a guess that as well as basic 'Is it poison or not?' gatekeeping, our taste buds also give us an appreciation of foods that are good for us, or, at least, were good for our ancestors. The availability of all these foods has, naturally enough, changed dramatically since the existence of early humans, which you can see as either a bad thing or a good thing depending on whether you've ever been responsible for the diet of a five-year-old.

Some foods, like fresh fruits for example, are bursting with wonderful perfumes even as they fall from the tree. Who could live without the scent of isobutyl formate, for example, better known as raspberry, or who wouldn't thrill to the warm tones of linalyl butyrate, also known as peach? This, as you might imagine, is the key to artificial flavours: work out what the key fragrances are in the food that you are trying to mimic, then manufacture them in bulk and add it to the thing you want to flavour. Of course, as far as your nose is concerned, the isobutyl formate in a real raspberry isn't any less isobutyl formate-ish than it is in a treated piece of gelatin. In our brave new world of food science, fragrances are just chemical compounds like anything else.

But back to cooking. Herbs, fruits, butter, milk: all of them have their own glorious odours. Yet when it comes to proteins like meat, eggs and flour, there really isn't much going on in the nasal department. Which, again, is one of the wonderful things about cooking; as we change the proteins into simpler, more digestible forms, we also generate a whole host of smaller, volatile, aromatic molecules that can stimulate our noses and therefore our palates. Because when it comes to enjoying what we eat, it's all about the spectacular collision of tastes and flavours, stimulating the brain in a unique, complex and utterly transporting way.

GOLDEN BROWN

So what kind of chemical reactions happen in food when we cook it? The answer, in a nutshell, is ones that involve sugars.

You can divide these sugar reactions into two basic groups. One of them you will already have heard of: caramelisation reactions. These occur when we heat food in a dry environment – such as an oven, grill or oiled frying pan – to high temperatures of around 160°C, and they result in the browning of the sugar as well as the release of a whole host of, well, caramel-flavoured aroma molecules.

The other kind of sugar reaction may be less familiar to you in name, but will be extremely familiar from the foods you eat. It's called the Maillard reaction, and it takes place at lower temperatures than caramelisation, usually kicking in around 140°C. Note, however, that's still well above the boiling point of water, and like caramelisation reactions it requires dry heat. It's essentially a reaction between proteins and sugars, turning them golden brown and releasing all sorts of extremely flavoursome aroma molecules. That glorious smell of freshly baked bread, or of bacon sizzling in the pan? Those fabulous aroma molecules come from Maillard reactions.[6]

When we bake a cake, of course, both these reactions may be going on, depending on the temperature that we bake at. If the temperature is over 140°C, the Maillard reactions will be happening at the surface of the cake mix, producing that familiar golden-brown crust. That's why brushing a little egg white

---

6 All of this, of course, goes a long way to explain why so many of your favourite foods taste so dreary when they are cooked in a microwave oven. The microwaves are tuned to the vibration frequency of water molecules, so the heat comes directly from the water in the food and not from hot air around it, or from hot metal, as it does in the case of baking or frying. As a result you don't get any wonderful browning Maillard or caramelisation reactions, or any of the tasty aroma molecules they bring with them.

on the surface of a biscuit helps it brown; you are just providing some extra protein for the carbohydrates – in the form of flour and sugar – to react with. And if the temperature is over 160°C, any sugars at the surface of the mix will start to caramelise and add other dimensions to the taste.

While we're on the subject of temperature, where Maillard reactions are concerned there's something else important to add: the higher the temperature, the less desirable the molecules that are produced. Most of the of bitter-tasting carcinogenic stuff gets made above 200°C, which isn't so relevant for cakes, perhaps, because we tend to bake at lower temperatures than we roast, but it is very relevant for meats. Muscle protein starts to denature (break down) at around 40°C, so the challenge of meat cookery with quality joints is really to get the inside of the meat above 40°C for long enough to cook it but not dry it, while getting the outside of the meat to a high enough temperature to kick-start the Maillard reactions.[7]

One tried and tested technique is to give the joint a blast in a hot oven or a frying pan, then turn the temperature down to cook it through. You'll hear many chefs tell you that this 'searing' is to help keep in the moisture while it cooks. The truth, however, is that searing makes no difference to the amount of water the meat retains. The water molecules are held in place by the proteins in the meat, and when the proteins start to denature the joint starts drying out. The reason that searing

---

7 Gristly meats, on the other hand, tend to have more collagen in them, and collagen turns to gelatin at around 50°C; this is why long slow stewing at lower temperatures works well for offcuts.

works is that it initiates the Maillard reactions at the surface of the joint, producing that delicious brown crust and releasing all sorts of wonderful meaty aromas.

A BLAST FROM THE PAST

The first day of your PhD is very different from your first day as an undergraduate. Your degree begins in a heady swirl of social engagements, timetabling and instant-coffee making; you are at the beginning of a journey that has been taken by thousands before you and will be taken by thousands after. Your task is to mould yourself to someone else's idea of how studying should be done, and to try to turn your interest to subjects and concepts that others have deemed worth while. You are, in short, a cog in a wheel, and that wheel is in turn part of a machinery that will spit you out in precisely nine terms' time whether you are ready or not.

Begin a PhD at the Cavendish Laboratory in Cambridge, however, and you are on your own. You have started a period of study that has no timetable; it will take as long as it takes, whether that be three years or ten. You are given a desk, you are told where the canteen is and you are introduced to the staff of the Cavendish Stores. By the way, if you are the type of person who enjoys a stroll around a do-it-yourself store of a Sunday, the Cavendish Stores are a strange kind of heaven. Not that you ever get to go inside; you stand in the queue at a small hatch waiting your turn, and the closest you get to all the goodies inside is a glimpse over the shoulder of the man who serves you. You can ask that man for anything, and he won't so much as blink: a three-phase power supply, a diode, a length of optic fibre – anything is yours for the sake of an awful lot of paperwork. Because your lab bench is empty and there is no equipment; the idea is that you will build any instruments you require from scratch.

As I stood in my kitchen a week before I was to meet Gordon Ramsay in a televised bake-off, I felt just as lost and hopeless as I did on that first day at the Cavendish. I had a good understanding of the science of baking, of course, but I had no equipment, no skill. And the first time I baked a Victoria sponge, from an old recipe handed down from my ex-grandmother-in-law, the worst possible thing happened: it was a complete success. I whizzed up all the ingredients in no particular order, greased an old cake tin, whacked it in a hot oven and out came the most glorious cake. It was as light as a robin's dream and as sweet as a milkmaid's kiss. I ate the whole thing pretty much single-handed, followed by a crippling sugar-rush. This was going to be easy.

It wasn't. Despite following exactly the same recipe and doing all the same things in the exact same order, the next three cakes were unmitigated disasters. The first collapsed as soon as it came out of the oven; the second was the texture of something you might use to clean a bike; and the third actually caught fire. Slowly my hubris curdled into shame. I was heading for a fall.

KARATE KID

My first year at the Cavendish was essentially spent as a miniature potter. After flapping around for a few weeks with little or no idea of how to fill in a Request Form for the stores, let alone how to direct my researches, I was assigned to the infinitely patient and sandpaper-dry-humoured David Hasko. David was kind enough to teach me a technique for making tiny gold patterns on semiconductor crystals. The goal of my PhD, you see, was to use these tiny gold patterns to make miniature electronic devices – devices so small that you would be able to see quantum mechanical effects.

The technique David showed me involved coating a semiconductor chip with a resist made of polymer film, patterning the film with an electron beam, then dissolving away the patterned part of the resist to leave a template. I would then coat the template with a thin layer of gold, dissolve away the remaining resist and, with a bit of luck, all that would be left would be the gold in the template. In case this is sounding a bit abstract, I've done a drawing. Draw a Christmas-tree shape with the electron beam, and at the end of the process you'd be left with a gold Christmas tree sitting on top of the chip. The point being, of course, that you could make very small Christmas trees indeed; I regularly made devices that were only hundredths of a millimetre across.

The theory was simple enough, but the sorry fact was that making a resist of exactly the right thickness was very tricky indeed. The method was to place the tiny chip on an only

How To Make a Tiny Gold Christmas Tree on a Semiconductor Chip

1/100 of a mm

Make this pattern with electron beam

Resist

Chip

Dissolve in Solvent

Christmas tree dissolves

Cover in gold

Gold Christmas tree sitting on chip

Chip

Dissolve in second solvent

slightly less tiny rotating mount a bit like a supercharged potter's wheel. You would then squirt some liquid polymer onto the chip, hit a footswitch and the chip would spin like crazy, spraying off most of the liquid resist but leaving a very thin, hopefully even film of resist clinging to the chip. You'd then hit stop and pop the chip in an oven to bake the resist solid, ready to be patterned by the electron beam.

There were a gazillion things that could go wrong. For a start, the whole thing was incredibly fiddly as the chip was far more likely to end up on the floor or in the turn-ups of your trousers[8] than on the frustratingly small chip mount. The size of the polymer drop that you put on the chip was crucial: a nanogram too much and the resist would be too thick for the electron beam to make it soluble; a nanogram too little and there wouldn't be enough height difference between the chip and the resist, and the gold wouldn't stick to the chip properly. How long you left the drop on the chip before you spun it, how fast you spun it, exactly where on the chip you placed the drop, the temperature of the room, the chip and the liquid polymer: everything made a massive difference.

The first six months of my PhD were spent following pretty much the same routine every day. Get some chips, spin some resist on them, bake them, measure the thickness of the resist, see if the patterning process worked. Occasionally, as with my successful Victoria sponge, I did a half-decent job, but the problem was that I never knew what I had done right.

8 No physicists wear lab coats by the way. Ever. Lab coats are for chemists.

Then finally, after hundreds of hours of practice, seemingly overnight I developed the knack. I knew just the amount of liquid to put in the syringe and exactly the number of split seconds to pause before I started the chip spinning. I could tell just from the colour of a resist how thick it was to the nearest thousandth of a millimetre, and whether it was worth trying to pattern it or whether it would be better to just wash it off and start again. I became the lab's go-to guy for little gold patterns on chips, and other more experienced research students in the lab began asking me to make devices for them in return for letting me in on the experiments they were running.

The point is, I guess, that spinning resist for a physics PhD and baking a Victoria sponge have a certain amount in common: they are what we call chaotic processes, where small changes in the initial state of a system can produce large changes in the end state. In both cases, we are dealing with the behaviour of large molecules; in the case of the resist, these molecules are relatively uniform in size and it's really just their physical properties – in other words, how they are affected by things like temperature, rest time and spin rate – that we are trying to get to grips with. In the case of a cake, things are even more complicated because we are dealing with a whole world of complex chemical processes in the form of Maillard reactions. And in the real world, we have only one way of dealing with chaotic processes: skill. I needed an expert and luckily Henry Dimbleby had one to hand.

A VERY DIFFERENT PTAK

Clare Ptak is a quiet legend in the world of baking. Originally from California, she was pastry chef at Alice Waters' restaurant Chez Panisse in Berkeley and now runs her own baking company, Violet Cakes, with a regular stall on Broadway Market in Hackney, East London. She also happens to be Henry's next-door neighbour, which was how she came to be a baking consultant for Leon and then his co-writer for the third Leon cookbook, *Leon Baking and Puddings*, which I heartily recommend to anyone who is at all serious about doing a bit of home cake-making.[9]

I thought the first thing Clare would do would be to pull apart my recipe, but she seemed very happy with it. The recipe I had for the sponge goes like this:

Weigh two eggs in their shells.

Weigh out the same amount each of butter, caster sugar and self-raising flour. That is to say, they should all individually be the same weight as the two eggs.

Cream the butter and sugar.

Add a few drops of vanilla extract mixed in with the two beaten eggs.

---

9 In fact, that book also contains a version of my Gordon Ramsay Victoria sponge recipe, together with an accompanying drawing I made of a man in a bath of meat.

Once the batter has been mixed thoroughly, quickly fold in the flour.

Place the mixture in a greased cake tin and bake at 180°C for about thirty minutes.

I have to say I remain suspicious that she just went along with the recipe because she didn't want to give me one of her own secret recipes, but I'll give her the benefit of the doubt. What she seemed much more interested in was the equipment I was using. My cake tin, she said, was completely unsuitable, being thick, square and intended for fruit cake; for sponge I needed a circular tin of very thin metal that would deliver the heat quickly and evenly from the oven to the cake. Next, she said that I wasn't even close to creaming the butter and sugar together for long enough. When she did it, she persisted for so long that the mix changed colour completely, losing its goopy yellow glow and becoming a stiff pristine white. This, she said, was the true secret of a really good Victoria sponge and, without it, despair would never be far away. I can't emphasise enough just how long she gunned away at that butter and sugar; if there was just one of her tips I could pass on, that would be it.

And finally, she said, I had to stop baking at such high temperatures. 'Slow and low' is the Ptak way, and by low I mean 140°C. That meant, of course, that the cooking time is an unusually dolorous forty-five minutes or so, sometimes even longer. And the test of the sponge being ready was that when the surface was prodded by a fingertip, it should just about 'remember' its old shape, and the sides of the cake should be pulling away slightly from the tin. Needless to say, her finishing and garnishing skills were exemplary; a dash of caster sugar and a couple of drops of vanilla in the whipped cream filling made a huge difference, and she insisted on using slices of fresh strawberry instead of relying, as I had, on industrial quantities of shop-bought jam. And she topped the whole thing off by tapping a little icing sugar through a sieve to give the surface of the cake a mouth-watering dusted effect. Gordon had better watch out.

HERE'S THE SCIENCE BIT

The theory of baking is relatively straightforward. Wheat flour consists of tiny capsules full of crystalline starch grains, and each capsule is coated with proteins. It's therefore a fabulous food, containing two of the major food groups, though in its raw form it's not as digestible as when it has been cooked. This, of course, is where good old bread and cakes come in.

The basic trick of baking is to add water to the flour to make a dough or mix, together with some agent that produces carbon dioxide when heated; the carbon dioxide then forms bubbles that expand and make the loaf or cake rise. In the case of bread, we tend to use yeast; in the case of cakes, we prefer what is known as baking powder.

Baking powder is made of sodium bicarbonate mixed with cream of tartar. The sodium bicarbonate releases carbon dioxide when heated, and the cream of tartar is there essentially to take away the bitter taste of the stuff that gets left behind. Self-raising flour is just plain flour that has already had baking powder added, to the ratio that you get about a teaspoon of baking powder to a cup of flour.

The amount of water compared to the amount of flour is what gives us the three main types of mix: pastes, doughs and batters.[10] From pastes we get pastries and biscuits; from doughs we get bread, scones and cakes; and from batters we get crêpes and pancakes, all depending on whether you add water neat, in milk, or in eggs, and whether or not you add fat and sugar. For example, biscuits are generally made from an egg/flour paste to which you add fat and sugar; scones are made from adding fat and sugar to a milk/egg/flour dough. In the cake recipe I was working from, typically for a sponge cake, all the water is in the form of egg with no other liquid added at all.

In the case of bread, of course, the dough is kneaded. Bread flour is high-protein flour, and kneading is a way of getting the proteins that coat the starch capsules within the flour to interact with one another and produce a tough new material that you will undoubtedly already have heard of: gluten. Baking the bread encourages the yeast to produce carbon dioxide bubbles,

---

10 The basic rules are these: for a paste, you need three times as much flour by weight as you do water; for doughs, twice as much flour as water; and for batters, equal quantities of both. Of course whether or not the water comes in the form of an egg (roughly 85 per cent water by weight), or as milk (around 90 per cent), or neat (100 per cent: just kidding) makes a huge difference to the resulting baked item, yielding all sorts of different taste and textures. Knowing this general principle, you can knock up pretty much anything you like in the kitchen without even looking at a recipe, although beware: doughs are much less forgiving than batters and pastes, and you will almost always need to get out your weighing scales.

which expand the gluten and make the bread rise. So long as the dough is baked at temperatures above 140°C, the complex Maillard reactions then begin, creating a wonderful brown crust and promoting the reaction of carbohydrates and sugars throughout the loaf.

Although bread relies on gluten for its structure and texture, a good light sponge should have as little gluten as possible. That's why in place of a protein-rich bread flour, sponge cake requires a protein-light flour. It's also the reason why, after creaming the butter and sugar and adding the eggs, the recipe calls for the flour to be 'folded' in; this is to encourage as little interaction between the flour proteins as possible and so avoid the formation of too much gluten.[11]

Yet for all my theory, I had still managed to make some dreadful sponge cakes, because sponge cakes, as luck would have it, are the exception to the general rules for how to raise bread and cakes. The main mistake I had made with my prototype pre-Ptak sponge cakes was in assuming that all the raising of the cake would be done by the carbon dioxide in the baking powder in the self-raising flour, so I didn't pay enough attention to creaming the butter and sugar. In fact, as Clare had shown me, this was a crucial step because its purpose was to trap as many tiny air bubbles as possible within the fat, which could then expand when heated in the oven. No doubt the baking powder in the self-raising flour does play a part, and it would be a brave man or woman indeed

---

11 Of course, this same principle of 'leave the flour alone' applies equally well in batters and biscuits if you don't want them to get tough and glutinous.

who attempted a sponge cake with plain flour. Clare had left me in no doubt: success in my forthcoming showdown with Gordon would depend entirely on the quality of my butter/sugar cream.

THE MILLER DIET

Before I recount that epic battle blow for blow, I think it's worth taking a brief diversion into the one area of food science that everyone seems permanently engaged with: calories. After all, cakes aren't known for their health-giving properties, and you may be concerned that I am squandering a chapter on proteins and the like without a whistle-stop tour through superfoods, sushi and other things that appear much more healthful and worthy. So for those seeking some nutritional advice beyond the usual eat-your-greens-healthy-balanced-diet, here is the best I can offer.

Do you count calories using the information on the back of food packets? Ever wondered how those values are calculated? Well, you may be surprised to discover it's simply by drying the food and burning it to see how much heat energy is released. All you need is a rather ingenious bit of kit called a bomb calorimeter. In essence, it's a little metal box suspended in water. You put some dried food in the box, set fire to it, then work out how much energy was produced by measuring the change of temperature in the water. When we say that a hazelnut has '9 calories', it is based on the fact that, if you dried it, stuck it in a bomb calorimeter and burned it, it would produce 9 kcal of heat energy.

Setting fire to a bit of food is chemically similar to digesting it, but not identical, so beware, because these calorific values work only as a rough guide. A raw potato, for example, might have the same calorie content as two Hobnob biscuits,[12] but your body is going to have a harder time extracting the calories from the raw potato than the processed Hobnobs and will therefore end up with more energy, once it has digested the biscuits, than it would after the raw tuber. Not only that, of course, but it's not much fun dunking a potato in a cup of tea.

What we tend to call 'bad' foods – things like cakes, hamburgers and sugary drinks – are really just foods that our bodies are very efficient at converting into energy. They may taste great, but the downside is that we don't have to eat much of them before we have ingested more than enough calories to get through the day. And as soon as you are producing more energy from your food than you are expending through living, you will start to store that energy as fat. If you are anything like a normal human being with a normal job, there is every chance that at some point in your life you will put on a bit of weight. Perhaps you think you are a bit overweight right now. If that is the case, before you fall prey to one of the many diet programmes that crowd the bookshelves, please do yourself the favour of trying out the Miller Diet.

---

12 My calculation is based on a largish 175 g raw potato at 0.77 kcal per gram, and two Hobnobs at 67 kcal a pop. I make that 134 kcal for both the spud and the biccies.

Let me start by saying, of any diet that has worked for you in the past – be it the Atkins, the Hay, the Raw Food, the F-Plan or the Lemon Cleanse[13] – they all worked simply because you ended up with fewer calories. Usually the key to the diet is that the food you have to eat (vitamin drinks, kale, low-fat foods, low-carb foods: you name it) is so unpalatable that no matter how hungry you are, you simply aren't going to eat enough of it to have more calories going in than you have going out. Other, more complicated diets such as the Hay (where you eat only food groups in certain combinations) will help you lose weight because you can never find anything to eat that fits the criterion. The Raw Food diet works, of course, because you just can't digest raw food as well as you can the cooked stuff. Whatever the diet – and I really don't care what it is – if it works, it works by reducing the amount of calories your body gets. That's it. Full stop.

So here's my diet: eat less food. Or, to be more precise, use up more calories than you digest. If you don't want to eat less, go ahead, but make sure you buy a bike or walk to work instead of getting the bus. I know, I know; you have an hour-long commute to your office and if you rode there on a bike, as soon as you had arrived and had a shower you'd have to start riding home again. Moan about it all you like: we're all in exactly the same boat. But don't, please, go on a diet that involves eating just green peas or cutting out fat, carbohydrate,

---

13 One of my showbiz friends once told me he had discovered an amazing new diet where all you did was drink lemon juice and maple syrup. 'And it's incredible,' he said, 'it just completely cleans you out.' He lost a whole lot of weight, which he claimed was from the 'cleansing' qualities of the lemon juice and maple syrup. Imagine that – not eating anything at all and losing weight. Ingenious.

protein, vitamins, minerals or fibre, because we need that stuff to live. Without fat, for example, you won't be able to make brain tissue; without carbohydrate, you won't have anything to power your muscles; and without protein you won't have any muscles in the first place.

But enough. It's time for pudding.

THE F WORD

The first thing that I need to say about Gordon Ramsay is this: he is not Gordon Ramsay. At least, not the sweary, confrontational taskmaster that we have grown to love in his cookery programmes. The man I met was taller than I expected, but then most people are taller than I expect because I am continually shorter than I expect. He was also friendly, good-humoured and intelligent off screen, but on screen, I have to tell you, all that changed.

The second thing to note is he really is very good at cooking. It's one thing to hone your cake routine in the quirky privacy of your own kitchen, and quite another trying to do it while Gordon Ramsay is next to you, chopping, measuring and seasoning with the sort of manual precision you usually see only when servicemen assemble weaponry blindfold. As I attempted simply to set my *mise en place* my hands began shaking so much I had to make light of it with one of those 'Look, my hands are shaking' comments that do nothing to relieve one's gut-wrenching embarrassment. And then the barracking started.

Gordon doesn't play fair. As you cook he goes out of his way to intimidate you, cursing you for choosing such a namby-pamby recipe in the first place, asking a question about science and accusing you of not making it funny, then hooting over how slow and domestic your kitchen skills are. It was as much as I could do to remember that I was baking a cake and that salt water from girlish tears might ruin the flavour.

And lastly I have to confess that I think Gordon's creation was far, far superior to mine in imagination and presentation. Instead of a traditional two-layer sponge, he decided to bake one large rectangle of cake and then roll it up with a delicious Eton-mess type filling of sweet cream, meringue and fresh strawberries that frankly had me wondering why I had bothered to turn up. Luckily I was saved by the Women's Institute, for it turned out that the four judges who would be blind-tasting our cakes were a little more traditional in their approach, and Clare's sponge was, of course, exemplary. On camera, Gordon took their unanimous decision with characteristic bad grace, but off camera he was full of praise. 'You know what?' he said, after the cameras had stopped turning. 'You'd make a half-decent pastry chef.'

CHAPTER 7

THE END OF THE WORLD IS NIGH

WHATEVER HAPPENED TO GLOBAL WARMING, EH?

As I sit down to write this chapter, England is in the grip of a cold snap. About an hour ago, snow began to fall and there's now a winter wonderland outside my window, complete with three drunken middle-aged men trying to build a snowman. There are headlines in all the national newspapers warning of a 'deep freeze' in the weeks ahead, and the odds against a white Christmas are shortening by the hour. All the major UK airports are snowbound, an arctic wind is blowing from the east and there are Met Office warnings on the television news that weather conditions over the next twenty-four hours are likely to deteriorate even further. So whatever happened to global warming, eh?

Are we all doomed?

The question of whither the weather has dominated public debate over the last few years. Every advancing warm or cold front seems either to deny that any problem exists or to condemn us all to bake in a high-temperature oven of our own making. There have been ex-politicians, such as Al Gore, who have held great sway with apparently scientific graphs and persuasive arguments, convincing many people of an 'Inconvenient Truth': namely, that the planet is warmer than it has been at any time in the last 10,000 years and it's all caused by man-made carbon emissions. Equally, there have been other, equally persuasive ex-politicians such as Lord Lawson, whose Global Warming Policy Foundation has the stated aim of 'bringing reason, integrity, and balance' to the debate about climate change, with equally scientific-looking articles about how temperatures in the UK have been getting colder over the last decade, not hotter, and claiming that the warm events have been caused not by carbon emissions, but by some mysterious Pacific Ocean current called El Niño. So what's the truth? Is it getting hotter? And if the greenhouse effect is the culprit, can we do anything to stop it?

Not only are these some of the most vital questions that face us as a species, but to answer them we shall be touring through some truly breathtaking scenery: the endlessly fascinating science of the Earth and its atmosphere. The constant heating and cooling of our planet as it orbits the Sun creates spectacular waterfalls, oceans, glaciers and ice caps; abundant warm tropical oceans and broiling arctic seas; tumbling cathedrals of cloud and blistering winds and hurricanes. Yet everywhere the same basic physics is at work, and once you begin to understand it, your sense of wonder only increases. The world of quarks and leptons is mind-boggling, but forever abstract, because the regime of the very small will never be something you can reach out and touch. The realm of jet streams and ice storms, on the other hand, will always inspire a different kind of awe, because they shape every day of our lives, as they have those of our ancestors and will shape those of our future children.

THE BIG CHILL

You may not know it, but you are living right in the middle of an Ice Age. Specifically, it's called the Quaternary Ice Age, after the geological period we are considered to be living through. If you really want to be pedantic, and I do, we are in the Holocene epoch of the Quaternary period of the Cenozoic era of the Phanerozoic aeon. There have been only four aeons since the beginning of the Earth 4.6 billion years ago, with each one spanning roughly a billion years: the first, the Hadean, covers the time when the first rocks formed; the second, the Archean, was when cyanobacteria came on the scene, generating the first oxygen; the third, the Proterozoic, was when single-celled eukaryotic life first appeared; and the fourth, our home period, the Phanerozoic, has seen the Earth festooned with every kind of plant and animal you can imagine.

We think there have been at least five Ice Ages throughout the Earth's history. In between them, as far as we can tell, there has been no ice at the poles. The first, the Huronian, occurred towards the beginning of the Proterozoic and lasted about 300 million years; the second, the so-called 'Snowball Earth' or Cryogenian Ice Age, came towards the end of the same aeon,[1] and lasted about 100 million years. The other three have all come in our own Phanerozoic aeon: the Andean-Saharan came first, but was the runt of the litter, lasting a mere 30 million years. The second, the Karoo Ice Age, ended 260 million years ago after clocking up a respectable 100-million-year reign. Our own Ice Age, on the other hand, the Quaternary, started a mere 2.6 million years ago. If Ice Ages were children, it's barely out of nappies.

I know what you're thinking: if this is an Ice Age, bring it on. But Ice Ages don't involve the globe being frozen over from start to finish; they are made up of cold periods, or glacials, where the ice caps advance, and warm periods, or interglacials, where the ice retreats. Looking back over the last ten cycles, the glacials have lasted on average about 90,000 years and the interglacials about 10,000 years. And here's the scary bit: we are living in one of those interglacials. And here's the even scarier bit: the last glacial ended 12,000 years ago.

LACKING IN ATMOSPHERE

When I was a schoolboy, back in the 1970s, all anyone was worried about was the onset of the next Ice Age, or, more strictly speaking, the onset of the next glacial period. All throughout

---

1 So-called because the planet is thought to have frozen from pole to pole. Erk.

the 1950s and 1960s the average global temperature had been falling, and every harsh winter seemed to herald a return of the woolly mammoth. This twenty-year cooling trend, as we shall see later, is now much better understood, but at the time it was largely blamed – in the media, at least – on the testing of nuclear weapons. Now don't get me wrong: as well as dazzling you with the wonderful science of the Earth's atmosphere and oceans, I hope to convince you that carbon dioxide emissions have been warming the planet. But it's hard not to notice that if you swap the words 'Ice Age' for 'Global Warming', and 'nuclear testing' for 'carbon emissions', you get something strikingly similar to the flavour of the current debate about climate. It makes great pasta, but mankind has a worrying tendency to consider itself on the brink of an apocalypse of its own making.

THE CHANGING OF THE SEASONS

So why is there no longer any fuss over global cooling? The main reason is that, since the 1970s, we're fairly sure we've figured out what triggers[2] glacial periods; in fact, the answer has been around for a while, thanks to the work of an enterprising nineteenth-century Scotsman named James Croll.

---

2 I say 'trigger' because once glaciation starts, it tends to get worse all of its own accord. The reason for this is something that scientists call 'positive feedback'. Once it's cold enough at the poles, you get an ice cap. Since snow is white, it reflects radiation, so more of the Sun's energy bounces off the Earth, making it cool further and causing the ice cap to grow, which reflects even more radiation, which cools the planet still further, which … Well, you get the picture.

Croll was a poor crofter's son from Perthshire who left school at thirteen and drifted through a series of unsuccessful careers as a farmhand, tea merchant, insurance salesman and hotelier before taking a job as a janitor in the Anderson Museum in Glasgow so that he could be nearer his one true love: books on geology and astrophysics. Completely self-taught, he developed an audacious and at the time completely unprovable theory: that glacial periods were the result of extremely slow changes in the orbit and angle of tilt of the Earth.

Now as you probably know, the Earth goes round the Sun in a roughly circular orbit, making one complete circuit in a calendar year. As it travels it spins on its own axis, turning near enough 365 times in one orbit. One of the things that make life interesting on Earth is that the Earth's axis doesn't point straight up; it's tilted at an angle, just as you see on every model globe. It's this tilt that gives us the seasons. Put simply, when the North Pole is tilted towards the Sun, the northern hemisphere experiences glorious summer; when it's tilted away, we get winter.[3]

In essence, Croll's theory was simple. The angle of tilt, the shape of the Earth's orbit, the direction of the Earth's rotation axis: none of them is constant. Each has its own period of variation, lasting several thousands of years. Take the shape of the Earth's orbit. Over a period of 100,000 years it slowly changes from a circle into something more like an ellipse and back

---

3 These are the solstices, from the Latin for 'still sun', as at these times of year the Sun is either northernmost or southernmost with respect to the equator. When the tilt is at right angles to the Sun, day and night are the same length in the northern and southern hemispheres, and you get the spring and autumn equinoxes or 'equal nights'. OK, I'll stop showing off.

again. The Earth also wobbles, like a top slowing down, with the tip of its axis tracing out a circle; one full circuit, or 'precession', takes about 26,000 years. Likewise the Earth's angle of tilt varies by a couple of degrees and back every 41,000 years. These gradual cycles change the contrast between the seasons; the colder the summer, the greater the chance of glaciation, because the ice never gets a chance to melt.

Right now, at the beginning of the twenty-first century, it looks as if we are a good distance away from the next glacial. The Earth's angle of tilt is unfavourable, being about halfway between extremes. Its orbit isn't helping either, being pretty much as circular as it gets. Only the Earth's wobble is doing the right thing. One of the effects of the wobble is to make the seasons occur at a slightly different point in the Earth's orbit from year to year. Right now, summer in the northern hemisphere happens when the Earth is furthest from the Sun, making it colder and thus improving the chances of glaciation.[4] To really trigger a cold snap, though, all of these various cycles need to be acting in step with one another, with each of them making summer cooler in one of the hemispheres. The best calculations we have don't predict a gangbang like that for another 23,000 years. As far as Ice Ages go, we lucked out; we began our civilisation in one of the longest interglacials in the last million years.

4 In actual fact, the Earth is closest to the Sun on 3 January and furthest away on 4 July. The difference in distance isn't enormous because the Earth's orbit is currently very circular: 152 million kilometres at the furthest point or aphelion, and 147 million kilometres at the nearest point or perihelion. As a result, the planet receives just 6 per cent more solar energy in January than it does in July, and summer in the northern hemisphere is just a little colder than summer in the southern hemisphere. Were the Earth's orbit at its most elliptical, that difference would be something of the order of 20–30 per cent, making our summers much cooler.

GREENHOUSE SCHMEENHOUSE

So given that as a species we are a terrible bunch of drama queens when it comes to the future of our glorious Eden, what can we be certain of when it comes to climate change? In other words, if the science of climate were a game show, what would be the prizes we would definitely be taking home? Well, if the first is that we are living in the middle of an interglacial in an Ice Age, the second is that carbon dioxide causes warming.

How can we be so sure? Well, the greenhouse effect, as it's also called, is vital to the survival of life as we know it. Put most simply, the atmosphere of our planet acts like a blanket, trapping the Sun's energy and raising the temperature of the Earth's surface. Without it, the average global temperature would be about −18°C. The average temperature of the Earth's surface is, in fact, about 15°C: in other words, the atmosphere provides about 33°C of warming that otherwise wouldn't be there. Funnily enough, the main greenhouse gas isn't carbon dioxide, but water in the form of vapour; about 21°C of the observed 33°C warming is due to moisture. Of the rest, about 10°C is caused by carbon dioxide. The rump – about 2°C – is due to methane, nitrous oxide, ozone and nasty industrial waste products like chlorofluoromethanes. The other main gases that make up the air – oxygen, nitrogen and argon – don't really figure as, unlike the other gases already mentioned, they don't have a warming effect.

The point to make here is that, when it comes to warming the atmosphere, carbon dioxide punches well above its weight. The amounts of carbon dioxide in the air are comparatively tiny – just

388 parts per million at the close of 2010. That's 0.0388 per cent, or roughly 100 times less carbon dioxide than there is water vapour, though as a whole carbon dioxide is responsible for nearly a third of the entire 33°C warming. What does this mean? It means that the temperature of the planet is very sensitive to the amounts of carbon dioxide in the atmosphere, that's what.

Thanks to the foresight of a man called Charles Keeling, we have a very accurate record of the amount of carbon dioxide in the air, dating back as far as 1958. That year, the carbon dioxide concentration, as recorded at the Mauna Loa Observatory in Hawaii, was 315 parts per million and, every year since, there's been a reliable 0.4 per cent annual increase in the amount of atmospheric carbon dioxide, as a result of the burning of fossil fuels. But what was the level of carbon dioxide in the more distant past?

Funnily enough, we have a pretty good record of atmospheric carbon dioxide, stretching back some 800 million years. As you now know, we are living in an interglacial at the beginning of the Quaternary Ice Age, characterised – like all ice ages – by polar ice caps. When snow settles it traps tiny pockets of air. As more snow falls on top, the older snow gets gradually more compacted, eventually turning to ice. By drilling down into the ancient layers of ice in Greenland and the Antarctic, and examining the content of the air bubbles trapped in the various strata, scientists are able to obtain a snapshot of the composition of the atmosphere through time. One such 'ice core', from Vostok in Antarctica, shows the concentration of atmospheric carbon dioxide varying between a minimum of 180 parts per million during glaciations and a maximum of 300 parts per million during interglacials. Why does the level of carbon dioxide change as the planet warms? Probably because of the varying solubility of carbon dioxide in warm and cold water; as you well know, warmed Coke goes flat. As the seas warm they release carbon dioxide, and as they cool they reabsorb it. The level of carbon dioxide in the present interglacial, the Holocene, has been fairly rock steady at 280 parts per million, until the 1850s when it started to hare upwards. The conclusion: the present levels of carbon dioxide in the air are the highest they have been for hundreds of millions of years, because of the burning of fossil fuels.

IS IT GETTING HOT IN HERE?

So we've won another prize on the Climate Change Game Show: levels of carbon dioxide in the atmosphere have increased by about 36 per cent since 1850. The third gift we're definitely taking home is that over that same period the temperature has gone up. In point of fact, over that same period of time the average global temperature has gone up by 0.75 degrees.

You can see this quite clearly in the graph on page 207, reproduced from the latest report of the Intergovernmental Panel on Climate Change (IPCC).[5] As you can see, it doesn't show

5 I should point out that the graph only goes up to 2007 and so doesn't show that the rate of increase of the warming has slowed down since 2006. By the same token, however, what the graph does show is that the rate of warming has slowed before, only to pick up again.

absolute temperature, but a temperature anomaly, plotted against time. The horizontal axis shows increasing years, starting with 1850 and ending with 2007, which was when the report was produced. The vertical axis shows the range of the anomaly, which varies from −0.9°C at the bottom of the vertical axis to +0.5°C at the top. What's an anomaly? Well, it represents the change from some long-term value. Let's say that, between 1980 and 2010, the average height for a man was five feet nine inches. If we measured the height of all men at the end of 2011 and found the new average height was five feet ten inches, then that's an anomaly of +1 inch when compared with the mean height for the previous thirty years. If the we measured men's

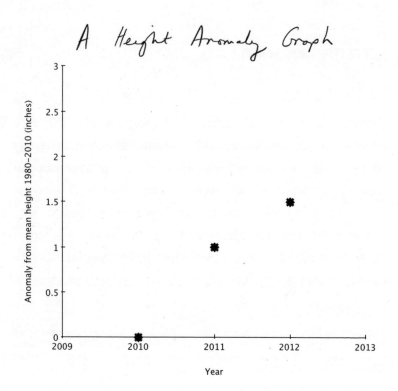

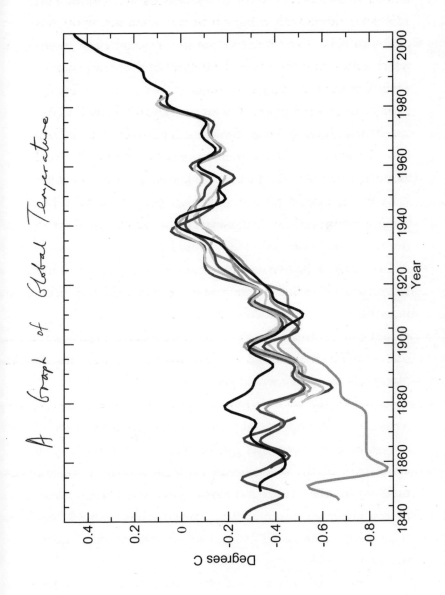

A Graph of Global Temperature

heights again this year and found the average to be five feet ten and a half inches, then that's an anomaly of +1.5 inches. Let's plot those two points on a graph and see what we get.

Right. So we have two data points showing height anomalies for 2011 and 2012, and we can say with certainty that the average height for men has increased by one and a half inches. Now here's the clever bit. Because we are talking only about anomalies, and not about absolute values, we have ruled out any inbuilt errors in the way the heights were measured in the first place. Say, for example, the person measuring the heights had a dodgy tape measure, and it's missing the first inch. When the real height was four feet eleven inches, they would measure it as five feet exactly. They would calculate the 1981 to 2010 average to be five feet ten inches, with an average of five feet eleven inches in 2011 and five feet eleven and a half inches in 2012. But when we come to calculate the anomalies, we would get the same answer as if they had a correct tape measure: +1.0 inch in 2011 and +1.5 inches in 2012, and when we plotted them out we would get the same graph.

It's exactly the same for temperature measurements. We can have a lot of confidence in this IPCC plot, because it shows not absolute temperature, but temperature anomaly, when compared with the average temperature of the years from 1961 to 1990. In other words, it doesn't matter whether someone's thermometer was reading 19.1°C when it was really 19.2°C; what we are comparing is the change in temperature against the 1961 to 1990 average.

What can give us additional confidence is that these measurements were made in all sorts of different ways. Some of the lines represent land stations and ship reports in the northern hemisphere only. Still others represent global land stations. Others use global land and sea surface temperature. Yet all of them, as we can see, tell us the same story: an increase of about +0.75°C, from −0.35°C in 1850 to +0.4°C now.

As you can see from the graph, the grouping of the lines gets tighter after 1900, as more and more weather stations were added to the various grids; this is a good sign, as you would hope that more measurements make for more accuracy and the various averages appear to be converging upon one another. I would go so far as to say that whatever it is that you're sceptical about with regard to global warming, it shouldn't be the increase in average global temperature. Although all these measurements were made with different instruments at varying times and places, they are all telling us the same story: since 1850 the average temperature of the Earth's surface has been increasing.

So point proven, right? After all, carbon dioxide is a greenhouse gas. The carbon dioxide in the atmosphere has gone up and the temperature has gone up; therefore surely the carbon dioxide has caused the increase in temperature. Well, not quite. It could just be coincidence. What if the extra warming due to the carbon dioxide from fossil fuels is being masked by some other as-yet-unknown cooling process, like an increase in the amount of cloud cover, and what if the temperature increase we're seeing is due to something else, like an increase in the amount of energy coming from the Sun? It's time to talk about the Medieval Warm Period and the Little Ice Age.

THE LAST FROST FAIR

However you look at it, the English winter of 1813–14 was on the nippy side. How do we know? Well, for one thing, England has the longest-running instrumental temperature records anywhere in the world, with monthly averages kicking off in 1659 and a daily record from 1772, so we can just look up the numbers.[6] And second, because London played host to a Frost Fair on the River Thames.

The Central England Temperature (CET) record is one of those glorious things that makes you proud to be a scientist. Covering a roughly triangular region, with Bristol, Lancashire and London at the corners, and spanning three and a half centuries, it is a hymn to measurement. While Napoleon was regrouping after his October 1813 defeat at the Battle of Leipzig, his much despised little shopkeepers in Ludlow were tapping their thermometers and recording the ambient temperature with their quills on their best parchment to the nearest tenth of a degree. Thanks to their efforts, we can see that after a relatively mild December, temperatures dipped just below freezing on 27 December 1813 and stayed there for a good month, with a particularly parky week around the middle of January 1814 when it never got above −4°C. By

---

6 If you're interested, you can even download them from the Met Office website at http://hadobs.metoffice.com/hadcet/data/download.html. The average daily temperatures are in the file 'Daily HadCET 1772–2009'. The units in this file are in tenths of a degree Centigrade, so don't be too impressed (as I was) if you see a −32 on 30 December 1813; that just means −3.2°C.

the end of January the Thames had frozen over upstream of the Old London Bridge, all along what we today call Bankside but which was then more commonly known as Queenhithe.

Although *The Times* declared the ice to be 'totally unfit for amusement' due to its 'roughness and inequalities', that wasn't enough to stop a gaggle of enterprising merchants setting up beer tents, amusements and souvenir stalls, and encouraging several plucky Londoners to venture forth. By 1 February the party was in full swing and an elephant, no less, was led across the frozen river to demonstrate to anyone who was still in any doubt that the ice was capable of bearing their weight. Luke Clenell's contemporary etching dated 4 February 1814, now in the British Museum, shows scores of punters skating, riding swingboats, cavorting and generally getting up to no good, which was all the more alarming as the festivities are recorded as having ended the next day and so the thaw may well have already been setting in.

The idea of the Thames freezing over by the South Bank seems bizarre in the context of London's current long-running spate of mild winters, and is only partially explained by the fact that at the time London Bridge was of a very different design, with nineteen narrow arches instead of the present five, which would have slowed the river and made it more vulnerable to freezing. Clearly there is something else going on here, and that something is what's known as the Little Ice Age.

BLOWING HOT AND COLD

The fact is, the sight of a frozen Thames was not all that star-tling to a Napoleonic Londoner. Things had been decidedly chilly for several centuries, with Norse settlers abandoning an increasingly nippy Greenland sometime around 1400, and reports of Henry VIII using the frozen Thames as a convenient short cut from London to Greenwich in the mid-sixteenth century. There had been Frost Fairs at regular intervals since 1608, with a particularly well-attended one in England's cold-est recorded winter of 1683,[7] when an entire street of shops was built across the river and entertainments included a printing press, puppet shows and a brothel.

The temperatures weren't, on average, anything like as cold as the last glaciation, which was some ten degrees or so below average temperatures today. But it was colder than it is now, on average, by about one degree. There is some debate about how global the cooling was; was it just a northern hemisphere thing, or did the southern hemisphere cool as well? Part of the problem here, of course, is we don't have a long-running instrumental record for all parts of the planet, but second-hand

7 The CET has only a monthly mean for the second half of the seventeenth cen-tury, with lower confidence limits than for the later daily series, as it includes non-instrumental data; a quick glance at the numbers for the winter of 1683 shows 0.5°C for December, then −3.0°C for January and −1.0°C for February 1684. That averages out to −1.2°C with a confidence of 0.5°C. So to make a comparison, the recent cold London winter of 2009 averaged 3.5°C with a con-fidence of 0.1°C. But don't take my word for it: look up the numbers at http://hadobs.metoffice.com/hadcet/data/download.html and get out your old school calculator.

evidence from glaciers in the southern Alps of New Zealand, for example, would seem to suggest that the cooling was reasonably widespread, cramping the style of a few other peoples beside those of northern Europe.

Whatever the truth of the matter – and if we're learning anything here, it's that thermometers are really handy when it comes to arguing about how hot or cold it is – you don't need to look at too many Bruegel paintings of whippets mucking about on frozen lakes to realise that the Little Ice Age was a big deal, at the very least, in most parts of the northern hemisphere. Talk of parties on the frozen Thames makes the whole thing sound positively entertaining, but the bad weather brought its share of misery too. The winter of 1815–16, two years after the last Frost Fair, is a case in point. The Thames may not have frozen over, but pretty much everything else did. In New England, the snow and ice finally departed in mid-May, only to return with a vengeance in June. The year 1816 became known as 'The Year without a Summer', or 'Eighteen Hundred and Froze to Death', with widespread crop failures causing one of the worst famines in European history, accompanied by mass outbreaks of typhus in the British Isles and bubonic plague in the Mediterranean.

Rolling back the years still further, the picture changes once again: this time, to one of relative warmth. We know that the Norse took advantage of retreating Arctic ice to colonise Greenland in the 980s, and that the five centuries from the year 900 to the onset of the Little Ice Age were at least as warm as the present day. There were cold snaps, of course – just as there were warm interludes in the Little Ice Age – but by and large the interval known as the Medieval

Warm Period in northern Europe was characterised by a growing population, fed on abundant harvests, who spent their surplus wealth and manpower building some extremely impressive cathedrals. Again, it is uncertain to what extent the heady days of 900 to 1400 in Europe were typical of what was going on across the planet as there is no direct temperature record; nevertheless, we have some good examples of what climate scientists call 'proxies', such as fossilised tree rings,[8] which indicate that, like the Little Ice Age, the Medieval Warm Period was more or less global in extent.

So we have a Medieval Warm Period from 900 to 1400 when global temperatures were on average at least as warm as they are today, and a Little Ice Age from 1400 to 1850 when they were about a degree colder. Of the two, we can be a bit more certain about the Little Ice Age because mankind was up and running with the thermometer by 1650 and making some pretty nifty measurements. We know from measurements of ice cores that throughout both these epochs the level of carbon dioxide was fairly constant at roughly 300 parts per million, or 0.03 per cent, so these shifts in temperature are unlikely to have been caused by greenhouse gases. So what was the cause? It's time to really get down to the nitty gritty and talk about the other two main drivers of the Earth's climate: sunspots and volcanoes.

---

8 Temperature by proxy is a fascinating business. The idea with tree rings is that the warmer the year, the more the tree grows. By measuring the thickness of the tree rings in years when the temperature is known, you can get an idea of what the temperature was in earlier years when there were no thermometers to record it. If you're smart you can cross-check with other sets of tree rings to get an idea of the margin of error, as well as with other proxies such as ice cores to make sure that you're on the right track. It's not ideal, but it's better than looking in Samuel Pepys' diary and reading that it was 'a bit chilly out'.

HERE COMES THE SUN

Let's take a step back for a moment and remind ourselves of the big picture. The Earth, our beautiful blue planet, revolves around the Sun in a slightly elliptical orbit. When the Earth is at its greatest distance from the Sun, it receives less of the Sun's energy than when it is at its closest. Since it is summer in the northern hemisphere when we are furthest from the Sun, this has the effect of making our summers cooler than they would be if they happened when we were closest. If the Earth's tilt was less, in summer we'd receive even less of the Sun's radiation and our summers would be colder still. And if the orbit were more elliptical than it is, we'd be further from the Sun and they'd be positively unpleasant, kick-starting a glaciation.

We've all got an intuitive feel for this; after all, how far away from the fire you sit clearly makes a difference to how warm you feel. But what about the heat of the fire itself? What happens if we throw another log on, or, indeed, if we don't bother and let the fire burn low? In other words, how is the Earth's climate affected by changes in the heat of the Sun?

Up until the 1980s it was assumed that the total energy radiated by the Sun was a constant, but since then we've had satellites measuring the energy it gives out on a daily basis, and they paint a fascinating picture. In fact, our home star blows ever so slightly hot and cold with a roughly eleven-year cycle, with a peak-to-trough variation of about 0.1 per cent. All in all, we've accurately measured the Sun's output through three of these cycles, and the last of them appears to be cooler than the rest, not by much, but definitely cooler.[9] Even more interestingly, the satellite data show a strong link between the Sun's radiation and the number of sunspots on its surface. And thanks to the efforts of a particularly enthusiastic amateur astronomer named Heinrich Schwabe, that means we can have a pretty good stab at reconstructing the Sun's brightness all the way back to 1825 and beyond.

SUNSPOTTING

Why 1825? Well, that was the year that Heinrich won a telescope in a lottery. He quickly developed a passion for stargazing. The following year he took delivery of a much more powerful refracting telescope from the Fraunhofer workshop in Munich, and turned the roof of his house into an observatory. Convinced that he might make his name as an astronomer by discovering a new planet, he began to make observations of the Sun, in the hope of seeing something career-defining cross its surface.

Now there are two planets between the Earth and the Sun: Mercury and Venus.[10] Both of them show up as small dots as

9 In fact, 0.02 per cent cooler. Just so as you know.

10 A transit of Venus is a must-see for astronomers. The last one was on 5 June 2012 and, as astronomical events go, caused quite a media storm. I hate to break it to you but Venus won't make another transit until 2117. By that time, if climate change has anything to do with it, we may have other things on our minds. Mercury is much more obliging as it is nearer the Sun and orbiting it much faster than the Earth: about four times as fast, in fact. It appears as a tiny dot, moving across the surface of the Sun over the space of a few hours, and its transit comes around on average every seven years or so.

they transit the face of the Sun, and Schwabe would have been looking for something similar from the roof of his house in Dessau. The problem is that the Sun's surface is often marred with dark markings called sunspots, and for Schwabe to be sure that he was seeing a new planet and not some temporary shadow on the Sun, he began to record the appearance of these sunspots in forensic detail.

His obsession grew, so much so that in 1829 he sold the family business in its entirety so that he could devote the whole of his working day to making his observations. A decade spent without so much as a sniff of a planet did nothing to dent his enthusiasm, at which point he made a discovery so significant that it would change everything we believed about the Sun. He found that the cycle of sunspots was beginning to repeat.

Of course, being the meticulous obsessive that he was, he didn't rush to print; instead he spent another ten years making more painstaking measurements to confirm his theory was correct, eventually publishing his findings in 1844. The world of astronomy was rocked on its axis. The Sun was no longer plastered with random pockmarks; there were hidden mechanics at work. Fellow astronomers soon came up with an answer: the equator of the Sun was rotating faster than the poles. The difference in speeds was causing a gradual warping of the solar magnetic field, until it could warp no more and broke out in a rash of surface activity, boiling up with immense solar flares, faculae (bright spots) and sunspots. Schwabe's work earned him the Royal Astronomical Society's prestigious gold medal. He had failed to find a planet but revealed the internal workings of a star.

THE SUN HAS CLIMATE TOO

Counting sunspots might seem like a pretty rough and ready way of working out what the Sun is doing, but when we match it up with the data we get from satellites we can see that it is remarkably accurate. The first person to make a detailed record of sunspots was Galileo, and if we include his observations we can extend our series of sunspot numbers all the way back to 1610.

What we can see on the graph is the year, plotted on the horizontal axis, from 1600 on the far left to 2015 on the far right. The vertical axis shows the number of sunspots, and you can clearly see the up-and-down variation due to the eleven-year cycle.[11] Even more fascinatingly, we can see two very significant minimums, where for extended periods of time there were a great deal fewer sunspots than normal; we call these the Maunder Minimum and the Dalton Minimum, and they occur slap bang in the coldest parts of the Little Ice Age. The Frost Fair of 1683, for example, pops up right in the middle of the Maunder Minimum, and that of 1813 makes an appearance plumb in the centre of the Dalton Minimum. We can also see that sunspot numbers build throughout the twentieth century to a peak

11 A contemporary of Schwabe, the Swiss astronomer Rudolph Wolf, compiled all the available data back to the cycle of 1755–66, which he labelled Solar Cycle 1. His data are marked with Xs on the graph. At the time of writing – winter 2011 – we are on the upsweep of Solar Cycle 24, which is expected to reach a maximum around May 2013. As that date approaches, expect your international flights to take a little longer, as aircraft will be avoiding the extra solar wind at the poles, so won't be taking any short cuts over the Arctic.

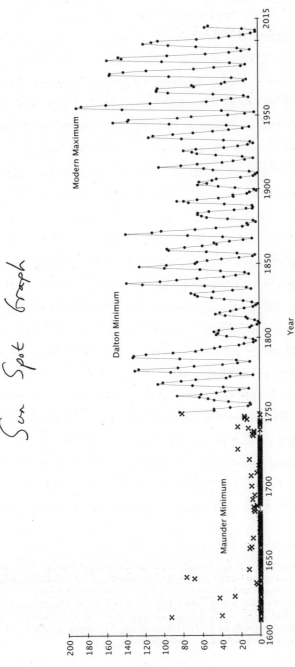

Sun Spot Graph

around 1950, dipping around 1970 before rising again to a lower peak around 1980. They fall away again between 1980 and the year 2010. We call these twin peaks the Modern Maximum.

We need to be cautious here: we are not measuring the energy emitted by the Sun; we are counting sunspots. Nevertheless, there's something to take note of. Sunspot numbers over this period mimic average global temperature, except for the thirty years since 1980. Take a look at the numbers for the twentieth century. There's a general build in sunspot numbers from 1900 to 1950, matching the rise in temperature we saw at the time. Global temperatures dropped a bit from 1950 to 1970 – that's why everyone was panicking then about global cooling – and we can see that sunspot activity declined in much the same way. Solar activity is not the be-all-and-end-all of climate change; as we shall see in a moment, there are several other equally fascinating factors at work. But though the Sun has undoubtedly been a player in the major climate events of the last 1,000 years, such as the Medieval Warm Period and the Little Ice Age, you can't say the same for the warming we've seen in the last thirty years. Over that period, the Sun has been getting cooler.[12]

---

12 In fact, using radioactive carbon dioxide in tree rings as a proxy, it is possible to reconstruct solar activity as far back as 10,000 years. More sunspots mean more solar wind, which means fewer cosmic rays, which means less radioactive carbon. When we look at an anomaly plotting of radioactive carbon concentrations in tree rings over the last 1,000 years, we can clearly see the Maunder Minimum, preceded by two others: the Sporer and Wolf Minimums. This pattern mimics the three known cold snaps during the Little Ice Age. Going back even further, we can see the Medieval Maximum, corresponding neatly with the Medieval Warm Period, and before that the Oort Minimum, when things were presumably rather chilly again.

ASH WEDNESDAY

On Wednesday 5 April, 1815, Mount Tambora on the island of Sumbawa in the East Indies detonated a colossal plume of dust into the blue sky above the Indian Ocean. The noise of the explosion was heard as far away as Jakarta on the neighbouring island of Java, some 1,200 kilometres away, where it was 'almost universally attributed to distant cannon' according to the British Lieutenant Governor, Sir Stamford Raffles. Nature had just fired a warning shot. Five days later, on the evening of 10 April, Tambora unleashed an astonishing torrent of molten rock and a giant column of noxious ash. One of the largest volcanic eruptions in recorded history had just begun.

In more or less a single sitting, Mount Tambora was reduced from 14,100 feet to just 9,350 feet, and the missing half-mountain of ash had been hurled some 40 kilometres up into the air, together with millions of tonnes of sulphur dioxide gas. As the heavier fragments came raining back down to earth, they created a baking-hot cloud of ground-hugging dust that rampaged across the island, killing some 10,000 people outright. The sky turned pitch black for a radius of some 600 kilometres, remaining dark for two whole days, and a tsunami up to four metres in height devastated the coastline of neighbouring islands. All in all, more than 60,000 people on Sumbawa, Lombok, Bali and East Java were estimated to have died as a result of the ensuing famine and disease.

VULCAN'S FORGE

Volcanoes are the wild cards of climate change and can have a dramatic effect on climate. The sulphur dioxide gas they give off, in particular, reacts with water vapour in the air to produce a haze of tiny droplets or *aerosol* of sulphuric acid, which then reflects sunlight at the top of the atmosphere, producing a cooling effect. In the case of Tambora, the devastation wreaked on the islands of the East Indian Ocean was just the beginning; the cooling effect of the ash and gas released, compounded by a cooler Sun, went a long way to producing the worldwide suffering of the infamous 'Year without a Summer'.

It's counter-intuitive, because somehow you'd think that an enormous mountain firing liquid rock into the air would warm things up a bit, but volcanoes actually reduce mean global temperatures. The dip in the world's thermometers in the 1950s and 1960s that got everyone panicking about an impending Ice Age was almost certainly caused by the activity of a number of contemporary volcanoes such as Mount Agung, which erupted in Bali in 1963, scattering over a cubic kilometre of ash into the upper atmosphere. Famously, the super-volcano Lake Toba, which erupted some 77,000 years ago on the island of Sumatra in present-day Indonesia, is thought to have released so much ash and gas that it triggered a global cooling of 3–5°C, wiping out most humans then alive.

If you were sceptical about what might be causing the current increase in global temperatures – and I'd say being sceptical is what science is all about – then you might wonder whether things are hotting up because of a lack of volcanoes. Well, again, it's interesting to note that there has been plenty of volcanic action over the last thirty years. The first half of the last century was maybe a little quiet, but since 1950 volcanic cooling has been vamping away solidly in the background and often stepping right into centre stage for a solo. We've already mentioned Mount Agung: 1980 brought us Mount St Helens; El Chichon blew in 1982, and Mount Pinatubo erupted in 1991, belching out a colossal 17 million tonnes of sulphur dioxide gas and lowering global temperatures in 1992 by 0.5°C.

Something, as we know, is warming the planet, but it's not a lack of attention from the great god Vulcan.

THOSE MAGNIFICENT MEN

So now we know the certainties of climate change. First of all, we are not in any imminent danger from a glaciation, other than from some unforeseen disaster like a super-volcanic eruption, a socking great meteorite impact, or someone in North Korea pressing the wrong button and accidentally unleashing thermonuclear Armageddon. Actually, come to think of it, none of that feels very reassuring, but heigh-ho. Second, the average surface temperature of the planet has increased.

We began by expressing that – as a lot of people do – by saying that there has been a 0.75°C increase since 1850. Knowing now that we were emerging from a Little Ice Age round about then, and that some of that warming was probably due to increased radiation from the Sun, let's focus on the roughly 0.5°C of temperature increase in the last thirty years. Throughout that time, the Sun has been ever so slightly dimming, volcanoes have had a cooling effect and quantities of carbon dioxide have been going up. So that's it, right? It's the carbon dioxide! Let's all get in our electric cars and go home! Again, we can't yet be that definite. We need to get a grip on the Earth's gloriously unpredictable weather.

On 15 March 1935, a one-eyed ex-barnstormer named Wiley Post took off from Burbank in California with the aim of flying east to Cleveland, Ohio. Post held the record for a round-the-world flight, having circled the globe in just eight days in 1931, then shaving a further six hours off his record in 1933. This time, however, Post wasn't aiming for distance. He was aiming for height.

The plane he was flying, a single-engine Lockheed Vega 5-C, didn't have a pressurised cabin, so Post had designed himself an extraordinary-looking suit, topped off by what can only be described as a deep-sea diving helmet. Connected to the suit was a bottle of liquid oxygen, which not only kept the suit inflated but also gave Post something to breathe. At an altitude of some 35,000 feet, he began to notice something very strange indeed. Although his aircraft's top speed was only 170 mph, he was achieving ground speeds of almost double that. He landed in Cleveland, some 2,035 miles away, just 7 hours and 19 minutes later: an average speed of 279 mph. Post had discovered the jet stream.

EIGHT MILES HIGH

You may not be a pioneer of aviation like Wiley Post, but thanks to your last international air flight you already know a great deal about the structure of the atmosphere without even realising it. You'll know, for example, that the weather is pretty much confined to the lower part of the flight, where the plane is climbing and bumping around a bit, and the buggers won't turn off the seat belt lights. Scientists call that region the troposphere, from the Greek *tropos*, or 'mixing'. The temperature in the cabin depends on how high they've set the air conditioning, but outside the plane it's getting colder; that's why you may have heard horror stories of stowaways on flights who bed down in the hold and end up freezing to death.

Then before you know it, you are up above the clouds and the plane levels off, becoming eerily still. You have reached the stratosphere, from the Greek for 'layer', and in look and feel it couldn't be more different. For a start, there's very little wind and next to no cloud, so the Sun beams in at your window as if it were the clearest, brightest summer day, even if you've just flown up through a pelting rainstorm on a grim morning in Brussels in mid-March. Commercial air travel really got going once the twin developments of pressurised air cabins and powerful jet engines made stratospheric flight a possibility. As well as it being nice and sunny, there was no longer any of that unpleasant lurching around you tend to get when making a trip in the troposphere. Flying in the stratosphere made people feel happier and safer, and that was great for the airline business.

Once the plane reaches the stratosphere, the seat belt sign usually goes off; the plane is rock steady, with clear visibility to every horizon, and can really start to cover some ground. If you're on a long-haul flight and travelling towards the east, then the chances are that at this point the pilot will navigate his way into the jet stream. First, he'll put the seat belt light back on, because there's a fair chance that, as the plane crosses the boundary between the relatively still air of the lower stratosphere and the high-speed winds of the jet stream, it might run into a little bit of turbulence. Then he'll steer the plane into a river of fast-moving wind a few hundred miles wide and a few miles thick, where he can basically get a piggyback that makes up for the forty-minute delay in getting everyone onto the blessed thing in the first place.

THE SPRING OF THE UNIVERSE

The jet streams are one of the most fascinating features of the atmosphere, and their existence gives major clues as to what drives the Earth's weather. There are two in each hemisphere: a polar jet and a subtropical jet. As the names suggest, the polar jets encircle the poles and the subtropical jets encircles what sailors call the 'horse latitudes'[13] at roughly 30 degrees above

---

13 Also the title of a great song by the Doors. Sailing ships would often get becalmed in the windless heat of the horse latitudes, and the sailors aboard them would slip slowly into a mindless torpor. 'Horse' is also a very cool name for heroin. Jim saw the link and went for it.

and below the equator. Astonishingly, like every other feature of the weather, their root cause is that the equator is hotter than the poles. That's right: rainfalls, snowfalls, tropical cyclones, the lot; they are all essentially cooling processes in fancy dress. Welcome to the wonderful world of thermal equilibrium.

To try to get our heads around what's meant by thermal equilibrium, let's forget about the atmosphere for a minute and take a bath together. Let's say that there's no hot water, so instead I fill my newly acquired, cast-iron, Victorian roll-top from the cold tap. Tragically, it's just a gnat's knucklebone above freezing temperature at 0°C and you quite rightly tell me I need my head tested if I think you're going anywhere near it. So I heat up a kettle of hot water at 100°C, and pour it in at one end of the bath. What's going to happen?

The answer, of course, is that you tell me the hot end is yours and the cold end is mine, and we both get in. Slowly, helped by my surreptitious wafting of the water, the water in the bath will start to mix, and soon the whole thing will be at the same temperature. Then slowly the temperature of the bathwater will start to lower until it is at the same temperature as the room, at which point you will ask me to go and fill another kettle.

So what's going on here? The answer is connected with something called the Zeroth law of thermodynamics, which, put in layman's terms, says that things like to be at the same temperature. Some pockets of hot water in the bath are less dense, so tend to rise, and other pockets are colder, so tend to sink, which creates one type of mixing called convection; but all the other eddies and currents caused by my secret wafting and convulsive shivering also have an effect, the net result being that all the water reaches the same temperature.

The bathwater as a whole is then hotter than the air in the room around it: pockets of warm air are created above the warm water, which rise and cool as they make contact with the air in the room; and the bath also radiates heat in the form of electromagnetic radiation, until the temperature of the bathwater and the temperature of the room are the same.

In case you think I've just taken a massive detour and all of this has been mildly diverting but ultimately pointless, let me reassure you by telling you that something very similar is going on in the atmosphere. The Sun heats the land and the oceans at the equator, but less at the poles, which causes a difference in temperature that the atmosphere and the oceans then attempt to even out via the weather. In a moment I'll explain how this works in detail, and how the jet stream and other wonderful things form as a result, but first it's worth saying a word or two about the predictability of bathwater and, by analogy, the climate.

THE BUTTERFLY EFFECT

Imagine for a moment that we were able to slow time down and look at the mixing of our bathwater. It would be extremely difficult to predict the precise eddies caused by my secret wafting of the water, say, or to know exactly what tiny pocket of warm water was going to rise to the surface at precisely what point as a result of convection. On a longer timescale, we know exactly what will happen: all the bathwater will reach the same temperature.[14] But on a short timescale, the exact temperature of, say, your toes, would be almost impossible to predict.

As it is with a bath, so it is with the atmosphere. We know what's going on: the various weather systems are attempting to even out the temperature difference between the equator and the poles. Nevertheless, predicting exactly how that happens on a short timescale is extremely difficult. Frantic wafting of the bathwater will sometimes produce only a tiny effect at your toes; at other times, the most subtle of hand movements will produce bath-long ripples that cause the water to slosh up over the top and onto the rug. The classic name for this, of course, is the 'Butterfly Effect', whereby the beat of a butterfly's wing in China can create a hurricane in Charleston. All things considered, the modern-day feat of predicting the weather with some confidence as many as five days in advance is an extraordinary achievement.

BIRD'S-EYE VIEW

Let's take a quick look at the Earth's weather from space so we can get the gist of how it operates. The first thing we see

14 Say there are nine parts cold water to one hot. Since the cold water is at 273 K, and the hot is at 383 K, the final temperature will be 283 K, or 10°C, provided no heat is lost by radiation.

is a belt of cloud around the equator and clear air at the poles. There's another belt of clear air at the horse latitudes; this, you will remember, is the location of the subtropical jet streams. Now let's focus on the northern hemisphere; above the northern horse latitude is a bewildering array of weather systems, all of them coming to an abrupt halt where they meet the polar jet. Looking down at the southern hemisphere, we see the same thing: all manner of weather systems south of the horse latitude, penned in by the southern polar jet stream.

It's easy to see what's going on: the Sun warms the land and the sea at the equator, and warm, damp, low-pressure air rises, creating the equatorial cloud. At the poles, cold dry air is sinking to form a low-precipitation area of high pressure. And in the middle latitudes, all hell is breaking loose as these warm and cold air masses mix. The jet streams themselves, we see now, are being caused where low-pressure warm tropical air meets high-pressure cold polar air, and they blow in a westerly direction because of the rotation of the Earth.[15]

15 Westerly means 'from the west'; in other words, the jet stream heads east. If we were to speed up the Earth's weather, so that we were watching the patterns of a year in just a few seconds, the multitude of weather systems in the mid-latitudes would just become a blur, and the idea of an average temperature for the planet would be instantly meaningful. But slow everything down to the speed of weather systems as they appear in real life, and the job of making predictions about temperature averages gets much harder. On a snowy December day in London, for example, that 15°C average is harder to get at. And therein, of course, lies the essential difference between weather and climate. Weather is what's happening outside the window right now; climate is its long-term average.

MOTION IN THE OCEAN

So far we've talked solely about the atmosphere, which is only human, after all, because we live at the surface of the Earth, breathing air and worrying about whether a sudden rainfall is going to make our hair go frizzy. But the troposphere is only half the story when it comes to spreading the Sun's energy out across the planet.

Ever since the Norse set sail across the Norwegian Sea at the end of the first millennium, bound for a newly melting Greenland, sailors have been aware of the existence of warm and cold currents in the oceans that could greatly speed their ships, in much the same way that modern airline pilots use the jet stream. The three main oceans on the planet, the Pacific, the Atlantic and the Indian, all carry outbound currents of warm water from the equator up to the poles, and return currents of cold water from the poles back down to the equator, in a mixing process that continually strives to bring them to a constant temperature.

The Indian Ocean, of course, is centred mainly in the southern hemisphere, so the majority of its currents shuttle between the equator and the Antarctic. The Atlantic Ocean has a bit of a bottleneck at the equator where the right hip of South America meets the left hip of Africa, which kind of cramps its style, but nevertheless manages to connect both poles and so provides substantial mixing. But when it comes to moving hot and cold water around and changing surface temperatures, there is only one ocean that is the daddy: the vast clear blue water of the Pacific.

THE NOT-SO-PEACEFUL PACIFIC

I've been talking as if the atmosphere and oceans were independent of one another, but of course they are highly connected, and nowhere is that more apparent than in the Pacific Ocean phenomenon of El Niño.[16] The western coast of South America is home to the famous Humboldt current, an upwelling of cold, nutrient-rich sea water that plays host to the largest marine ecosystem in the world, where 20 per cent of the world's fish is caught. Once every five years or so, however, the Humboldt current is disrupted by a warming of the waters of the east Pacific, with devastating consequences for the fisheries and affecting the weather of pretty much the entire planet. This disruption is called El Niño.

The warm waters cause excessive rainfall in the usually dry east Pacific, drought in the west Pacific and generally warmer temperatures worldwide; the large El Niño of 1998, for example, is the third warmest year on record, nestling just behind 2010 (the first) and 2005 (the second). Likewise, the opposite phase of the cycle, La Niña, causes warming of the western Pacific, with high rainfall in eastern Australia, drought on the west coast of South America and a drop in worldwide temperatures.

So, finally, could El Niño be causing the warming effect that we've seen over the last thirty years? The answer is probably

---

16 The name means 'the boy' in Spanish, and refers to the Christ child, since the warm waters off Chile and Peru are often first noticed around Christmas time.

not.[17] El Niño is an internal effect, an oscillation in a what-goes-up-must-come-down kind of way. Over time you'd expect as many El Niños as there are La Niñas, with the two effects cancelling each other out. A ramping increase in global temperature means a ramping increase of energy in the system, which means that our only remaining candidate, folks, is – you guessed it: carbon dioxide.

LET ME INTRODUCE YOU TO A MODEL

So carbon dioxide is very likely to have caused most of the 0.5°C warming of the last thirty years. Seeing as we're still all burning fossil fuel with merry abandon, what does that mean for the future? Well, thanks to ridiculous advances in computing power and ever more accurate information about the real climate from satellites and observation stations, we can do something quite extraordinary: we can build a virtual model of the Earth and run the numbers.

First of all, there's nothing that special about a computer model. It's just a way of collecting together all the things we know about the Earth's climate and then using that knowledge to try to predict how it will behave in the future. The Met

17 Not 'definitely not', because there's some evidence for a decadal variation of the El Niño/La Niña oscillation. In other words, there are decades with lots of El Niños and decades with lots of La Niñas. If that were the case, then it could be that that we have happened to have three El Niño decades in a row. Over a long enough period of time, of course, when we have sampled a great many decades, the La Niña phase will establish itself and it will all cancel out.

Office, for example, has the Unified Climate Model, which it uses to produce the UK weather forecast. They don't use all the bells and whistles when they are producing the five-day forecast for Rutland, but they do when they run simulations of the global climate decades into the future.

Our understanding of all the various processes that occur in the atmosphere and the oceans isn't complete, so the models aren't complete either. We've got a good grasp of most of what's going on; we can use mathematics to describe the behaviour of the atmosphere and the oceans, and how they interact with one another, but for other elements of the real world it's not quite so easy. One day, I'm sure, we will be able to write down the equation for a tree, but for the moment we have to measure the effects of forests on the atmosphere and make our best guess as to how to treat them in the model. Other things that you might think were straightforward to model are fiendishly complicated; clouds, for example. There are a squillion different types, for a start, and where they sit in the atmosphere makes a big difference as to whether they help cool things down or start to heat things up.[18]

Another limitation of the models, of course, is that we have no way of testing their long-term predictions about climate, other than waiting and seeing if they are right, which sort of defeats the purpose of a prediction in the first place. We know that most of them mimic recent trends in mean global

18 Stratocumulus clouds – those big white fluffy ones that you see in the lower troposphere – provide a net cooling effect by reflecting sunlight in the day and absorbing radiated heat at night. Your classic upper tropospheric cloud, on the other hand, the wispy cirrus, absorbs more than it reflects, thereby contributing a net warming.

temperature extremely well, but then some of them would, because they were calibrated using that information in the first place. We can give them a slightly stronger test by comparing their predictions of historic climate, but our knowledge of past climate is sketchy, so again it's hard to know how accurate they are. Nevertheless, I'd suggest it's worth knowing that an average of the top twenty climate models, compiled for the latest report of the Intergovernmental Panel for Climate Change, returned projections of a mean global temperature rise varying between 1.8°C and 4.0°C by 2100, depending on how drastically we curb our carbon emissions. That figure of 4.0°C, the upper limit, is a lot. Given a few centuries, that's hot enough to comfortably melt the Greenland ice sheet, adding six metres to the sea level, and that's before any melting of the Antarctic.

SO COME ON, ARE WE ALL GOING TO DIE OR NOT?

The answer, of course, is yes, at some point; whether or not we crash the car now, the road becomes something more like a cycle path in a few tens of thousands of years when we hit the next glacial, and it falls off a cliff when the Sun expands to form a red giant in about a billion years' time, vaporising the oceans and returning the planet to much the same state as it started in. And all along the way, at virtually any moment, our species could throw a seven by running into a super-volcano that plunges temperatures worldwide or a giant space rock that hits us with such an enormous impact that it disrupts the Earth's magnetic field, leaving us unprotected from the solar wind.

But in reply to the question of whether we can triumph in the face of global warming and buy ourselves and all foreseeable generations of our species a future on the most desirable planet in the solar system, I would have to say 'yes'. As we've seen in this chapter, our best guess is that carbon dioxide emissions from fossil fuels have been warming the planet and have the potential to change the climate considerably if they continue to rise at similar rates. Clearly we need to keep measuring, and debating, and trying to gain greater understanding of the processes that drive climate. But we aren't facing Armageddon; at least, not while we have the will to act.

We should always be sceptical about science. But being sceptical, to me, means open-minded and humble in the face of evidence. The Al Gores of this world, in my opinion, are too cocksure: there was a Medieval Warm Period; it's overstating the current warming to compare the temperature now to that at the end of the Little Ice Age; and carbon dioxide is one part of a complex system full of checks and balances, and unlikely to be the 'control knob' of global climate. On the other hand, if it isn't carbon dioxide that's causing the warming, what is it? Carbon dioxide may not be the only cause, but it sure as hell can't be helping.

We need to get to it. There's no need for hand-wringing and pessimism. The climate is changing, because it has always changed. Our species was a product of that change, being highly adapted to survive the low-temperature extremes of the last glacial, and our civilisation has thrived throughout the warmer temperatures of the Holocene. We have many emerging technologies up our sleeves such as nuclear fusion, which could provide us with almost limitless carbon-free energy by harnessing the same process that fuels the Sun, and carbon capture, which would enable continued burning of fossil fuels without releasing CO_2 into the atmosphere. And, being an optimist at heart, I can't help thinking that any number of equally ingenious solutions will suggest themselves as our knowledge of climate continues its breathtaking expansion. We are, after all, *Homo sapiens*. Climate change is what we do best.

CHAPTER 8

FLY ME TO THE ASTEROID!

THERE'S A STARMAN WAITING IN THE SKY

When I was five, if you'd asked me what I wanted to be when I grew up, I'd have immediately told you, 'An astronaut'. Come to think of it, if you'd asked me the same question at forty-five I'd have given the same answer. I am now more or less coming to terms with the possibility that I may never set foot on the Moon, though the fact that Alan Shepard commanded Apollo 14 and took lunar golf shots at the age of forty-seven still gives me some hope. Granted, he had previously been the first American in space and Chief Astronaut at NASA, whereas my experience is primarily in television sketch comedy, but enthusiasm has to count for something, surely?[1]

---

1 Shepard was also responsible for the Space Programme's finest joke when he said of his Mercury mission, 'They wanted to send a dog, but they decided that would be too cruel.'

Whatever my chances in the current round of NASA recruitment, the Moon landings will always hold a special place in my heart. As a child of '66, I was too young to watch Neil Armstrong and Buzz Aldrin take the very first moonwalk, but the following year I watched live as David Scott and James Irwin of Apollo 15 romped around the lunar surface in what looked very much like the buggy from my favourite TV show *The Banana Splits*. For me and a generation of children like me, the Moon landings were the very reason we wanted to study science. The Large Hadron Collider might take your breath away in its scale and ambition, and the control room of the JET (Joint European Torus) fusion laboratory in Oxford may feel like the bridge of a twenty-second-century spaceship,[2] but even those two towering achievements are dwarfed by the commitment, audacity and ingenuity of the Apollo missions.

Yet, astonishing as it may seem, there are a great number of educated people who actually believe that the Moon landings were a hoax. At least, that's what I'm guessing from the number of them who seem to end up sitting next to me at dinner parties. The gist of their argument, from what I can make out, is that the rockets and the spacesuits were just a

2 I was lucky enough to visit the JET laboratory in Oxford during the making of my BBC *Horizon* programme *One Degree*, and it's the closest thing I've ever experienced to stepping into the future. JET fuses hydrogen into helium while releasing energy, just like the Sun. The two main advantages of nuclear fusion over nuclear fission are first that the stuff you put in is much more readily available – hydrogen rather than uranium – and second that the waste products are radioactive for hundreds of years rather than thousands. Basically, if the human race wants to keep partying it's hard to think of anything that can keep the lights on other than nuclear fusion. So far the energy consumption of JET is greater than the energy given out by the fusion reactions, but JET's successor, ITER (International Thermonuclear Experimental Reactor), is designed to use less energy than it produces. Don't hold your breath, though; it doesn't come online until 2033.

front; the footage of Armstrong and Aldrin emerging from the lunar module was mocked up in a film studio and beamed around the Earth in order to convince the Russians that the race to the Moon was over.

Although I feel as if I am killing a tiny part of my soul in doing it, I'm going to take a look at the most common 'evidence' that NASA faked the lunar landings and, I hope, convince you that it did nothing of the kind. In fact, I hope to do more than that: to rekindle your sense of wonder at what was achieved in that vital first step of our quest to leave the Earth and populate the worlds beyond. And an understanding of the Apollo Moon landings is more important than ever, because this decade is about to see a step increase in space travel.[3] Indeed, if Richard Branson has his way, Virgin Galactic will soon be giving hundreds of paying customers their jollies 100 kilometres above the surface of the Earth, right at the very edge of outer space.

This book has been all about my favourite bits of science, but if I had been allowed to write only one chapter, this would have been the one I'd have chosen. Because as ever with science, the truth is far more bizarre than anything that the conspiracy theorists can make up. We really did land on the Moon, and what's more we did it using principles of motion and gravitation that were established in the seventeenth century. And,

3 China plans a Moon landing by 2020; Russia is working on a manned Mars mission and has just completed Mars500, a psychometric test for cosmonauts on a simulated Mars mission. Plus NASA is pursuing Barack Obama's new goal of a manned landing on an asteroid by 2025. The next date in my diary: 2014, when the European Space Agency's Rosetta probe will land on the comet 67P/Churyumov–Gerasimenko.

bizarre as it seems, some scientists really do believe that there are intelligent life forms on other planets; in fact, they have been making concerted efforts to contact them ever since the discovery of radio waves in the late nineteenth century. We'll be running the numbers on something called the Drake Equation for the number of planets in the Universe that may harbour detectable life, and asking what the chances are that we will make contact with an alien civilisation in our lifetimes. As we shall see, our future really does lie in the stars.

WHEN I GET YOU UP THERE

So when does the story of our trip to the Moon begin? One argument would place it with the Wright Brothers, the bicycle manufacturers from Dayton, Ohio, who made the first powered flight in the *Wright Flyer* in 1903. Neil Armstrong, for one, certainly seems to see himself as an aviation pioneer rather than a spaceman, and even took a piece of the *Wright Flyer*'s wooden propeller to the lunar surface on Apollo 11. Others might start even earlier with Sir George Cayley, the inventor of the first glider in 1849 and the founder of modern aeronautics.[4] The more poetic might place it with the wandering spirit that led *Homo sapiens* out of Africa some 70,000 years

---

4 Cayley defined the main forces acting on an aircraft to be lift, drag, thrust and weight. Drag acts opposite to the direction of flight; lift is perpendicular to the direction of flight; thrust acts in the direction of the engine; and weight acts straight down. For a spacecraft landing on the Moon, there is no drag and no lift, just thrust and weight.

ago. But for me, space travel begins with one man: the great Sir Isaac Newton.

STANDING ON THE SHOULDERS OF GIANTS

The world is a teeming confusion of objects in dizzying motion. Leaves romp in the breeze, clouds scud across the sky, and cars beetle along busy roads heaving with restless people on a planet that is spinning fast on its axis, in a ceaseless orbit around a white-hot nuclear inferno spewing off blinding light in every conceivable direction. We end each day battered and bruised, nursing stubbed toes and aching limbs, exhausted from standing upright in a strong gravitational field. Yet we are each able to make sense of this daily round of collisions, orbits and motion, thanks to one of science's true geniuses.

Isaac Newton was a truly remarkable man. A grammar-school boy from illiterate Lincolnshire farming stock, born on Christmas Day 1642 into an England gripped by Puritan fever and blighted by continual skirmishes between Parliamentarians and Royalists, he rose to hold the Lucasian Chair in Mathematics at Trinity College, Cambridge,[5] was President of the Royal Society, a Member of Parliament and a Master of the Royal Mint, and he was buried at Westminster Abbey. He was obsessive, secretive and emotionally cold but with a choleric

---

5 If this post seems familiar, it's because Stephen Hawking famously held it from 1979 to 2009.

temper, and his life's work is best summed up by the epitaph written by Alexander Pope:

> Nature and Nature's laws lay hid in night.
> God said, *Let Newton be!* and all was light.

Newton's achievements stand equal to, and possibly even surpass, those of the other true scientific genius, Albert Einstein. But whereas Albert possessed a genial, almost Buddhist calm and seemingly childlike self-confidence, Newton's gifts came in an altogether different package. A loner whose father died shortly after his birth and whose mother abandoned him for a new husband when he was three years old, Newton was riddled with insecurity and unable to bear criticism, publishing little that he wrote. Partly this was due to the warp of his personality; partly it was because of a natural desire to conceal his unconventional – and heretical – Christian beliefs;[6] and partly it was because he stood in the hermetic tradition of the adept and the sorcerer rather than that of the modern scientist. In fact, you'd be hard pushed, really, to call Newton a Newtonian in the modern sense of someone who believes in a clockwork Universe; instead he held that the natural laws he discovered were created by God and were compelling evidence of His design.

---

6 This sounds like a bad Dan Brown novel, but Newton is commonly thought to have been a member of the heretical Arian sect that had split from the orthodox Christian Church at the First Council of Nicaea in 325, and who rejected the idea of the Holy Trinity, believing instead that Christ was mortal. To my knowledge there is nothing to make the direct link, though we know from his writings that Newton was undoubtedly an antitrinitarian. Predictably, Newton is also variously believed to have been a Mason, a Rosicrucian, an initiate of the Priory of Sion and a fully paid-up member of the Magic Circle. All right, I made that last one up.

In his downtime from exhaustive researches in alchemy and his endless hours spent scrutinising the Bible for hidden messages and prophecies, Newton developed a theory of light and optics, creating one of the most vital tools in mathematics in the form of the calculus. But it was his theories of motion and gravitation that laid the foundation for, well, just about every achievement of science and engineering of the last three centuries. Today, they appear to us to be so simple as to be common sense, but they are a human invention. With three simple laws of motion and one of gravitation, Newton took a teeming confusion and resolved it into a Universe we could understand, mathematically describe and therefore truly manipulate.

YOU DON'T NEED A FORCE TO MOVE

The first of Newton's three laws of motion appears so straight-forward as to be hardly worth the ink, but to me at least it is one of the most subtle, far-reaching and downright beautiful pieces of prose in the English language. The more you ponder on it, the more there is to it. Here it is as Newton expressed it in his masterwork *Philosophiae Naturalis Principia Mathematica*, often called simply the *Principia*: 'Every body perseveres in its state of rest, or of uniform motion in a right line, unless it is compelled to change that state by forces impress'd thereon.'

Let's think about that for a moment. At first glance all it seems to say is that you need to push something to change its motion, which I think we'd all agree with. But there's much more to it than that. Newton is saying that motion in a straight line at constant speed doesn't require the action of a force. But how can something move without being pushed? That's crazy, surely?

Aristotle certainly thought so. He believed that an object moving in a straight line at a constant speed required a force to push it, and it's easy to see why. Here on the Earth, we are so used to working against friction and air resistance that it seems as if it requires effort to keep an object moving, but it doesn't, not if there's nothing to drag it to a halt. Think of striking a golf ball, for example. There are two main effects that limit the range of your shot. First, the ball has a limited flight time before gravity pulls it down to Earth, and second, air resistance will be working against it as soon as it leaves the face of your club. Third, there's also the fact that in my experience hitting a golf ball with a golf club is very nearly impossible, but we'll leave that out for now.

Play a few holes on the Moon, however, and you'll really start to love Newton's first law. Alan Shepard muffed his first shot, but his second went nearly a mile into the distance. With no air resistance working against it, the ball's horizontal speed remained constant at the speed at which it had left the face of Shepard's club, exactly as Newton predicted. The only thing that limited the range of the shot was the pull of the Moon's gravity.

Now think about swinging something on a string around your head; maybe like my five-year-old son you keep your mittens at either end of a piece of elastic and for devilment you like to pretend that one mitten is the blade of a helicopter. Here's an alarming and extremely counter-intuitive fact that comes straight from Newton's first law. To keep the mitten spinning in a circle at a constant rate, you just need to pull the mitten towards you. After all, according to Newton's first law, the mitten wants at any moment to keep moving in a straight line; in fact, if the elastic snapped, it would career off at a tangent to the circle in which you were spinning it.[7]

This has big implications for orbits. An orbit is just one mass circling another – the Moon around the Earth, for example – but you will have noticed that the Moon isn't attached to the Earth by a piece of elastic. Some other force is pulling the Moon around the Earth in a circle. In a breathtaking leap, Newton realised that the force that held planets in their orbits was one and the same with the one that made the apple fall straight down from the tree. And in a moment of uncharacteristic poetic fancy he gave it a fantastic name: gravity.[8]

TO ACCELERATE SOMETHING YOU NEED A FORCE

But we'll come to gravity properly in a moment. Let's stick with Newton's laws of motion, because they have a big pay-

7 OK, in practice, to keep the mitten moving in a circle you do need to supply a small tangential force to overcome air resistance, which you do by making small circles with your hand rather than holding the elastic stock-still. And, of course, you need to supply a big tangential force to get the thing moving in the first place. But the main point to grasp is that, once started, and in the absence of friction forces, circular motion requires only a radial force to maintain it.

8 At least, that's how it translates into English. Newton wrote in Latin, so he called it 'gravitas'. By the way, for sheer fun the translations I've used are the closest I can find to the written English of the time of Newton. They are by one Benjamin Motte in 1729, two years after Newton's death and around forty years after the original publication date of 1687. These days, of course, physicists state the laws in modern English and don't wear wigs.

off in terms of spacecraft. The Big Man's second law of motion reads like this: 'The alteration of motion is ever proportional to the motive force impress'd; and is made in the direction of the right line in which that force is impress'd.'

Beautiful, I know, but what does it mean? Well, Newton is saying that motion – as we now know from the first law – doesn't require the action of a force, but a change in motion does. A change in motion, of course, is what you or I would call acceleration. When we press the gas pedal in our clapped-out Ford Kia, the back wheels push harder against the tarmac and we go faster along the road. And if the back wheels double the force on the road, the Kia's acceleration will double, at least it will if the bodywork holds out.

If we express this law in mathematical terms, we get the most useful equation in the whole of science:

$$F - ma$$

Where F is the net force in Newtons, m is the body's mass in kg, and a is the body's acceleration in m/s^2. I know this sounds a bit like maths, but don't worry about that for now, because the important point to grasp is simply the fact that Newton has defined mass. In Newton's model, mass is a constant: an unchanging property of all bodies that resists changes in their motion. The more mass you have, the harder it is to make you accelerate, as anyone will know who has attempted a fireman's lift. We are all taught as much with our mother's milk, and it's simply stunning to think that something so fundamental to how we see the world was once one man's idea.

Interestingly, it turns out that Newton is wrong here, but he is wrong to such a minute degree regarding the masses and speeds we come across on Earth that no one really noticed the error until Einstein came along in the twentieth century. As Einstein showed, in fact a body's mass isn't constant. It depends on the speed it is travelling at, according to the equation

$$m - m_0 \frac{1}{\sqrt{1 - \left(\frac{v}{c}\right)^2}}$$

At low speeds, where the body's speed v is less than c, the speed of light, the body's mass m is pretty much equal to its rest mass m_0. As its speed approaches the speed of light, its mass approaches infinity. This has some serious implications for our chances of meeting aliens, because it implies that nothing travels faster than light.

FORCES ALWAYS COME IN PAIRS

So Newton's first law tells us that stuff, left to its own devices, likes to either remain at rest or move at constant speed in a straight line. The second law tells us that to speed stuff up we need to push it. And the third law tells us that we can't push something without it pushing back, or as Newton put it: 'To every Action there is always opposed an equal Reaction: or the mutual actions of two bodies upon each other are always equal, and directed to contrary parts.'

In other words, there no such thing as a single force; forces come about when stuff interacts with other stuff. Walk into a lamp-post in the street and you will get a nasty smack in the face, while the lamp-post gets an equal thump in return. It is this Law, as we shall see, that presently holds the key to manned space travel. But to travel from the Earth to the Moon, of course, we need to understand gravity. Once again, the theory that we use is that of Isaac Newton, formulated back in the mists of the seventeenth century.

STUFF ATTRACTS OTHER STUFF

'. . . there is a power of gravity tending to all bodies, proportional to the several quantities of matter they contain . . . the force of gravity . . . is reciprocally as the square of the distance of places from the centre of the planet.'

That's one way that Newton puts it in the *Principia*, and it's worth taking a moment to absorb exactly what he means. In fact it's probably easiest to see what's going on if we express this law mathematically:[9]

$$F - \frac{GMm}{d^2}$$

---

9 Just in case it's a while since you last flirted with algebra, this is just a gentle reminder that *GMm* means *G* multiplied by *M* multiplied by *m*. The equation is saying that the gravitational force *F* is equal to a fixed number *G* multiplied by the mass of the first body *M*, multiplied by the mass of the second body *m*, divided by the square of the distance between those two bodies *d*. 'What happened to multiplication signs?' you may ask. The answer is that we tend not to use them because they look too much like another favourite algebraic letter, *x*.

In this equation, F is the force between two bodies of mass, M and m, and separated by a distance, d. Don't worry about the G for now; that's just a fixed number.[10] The point to grasp is that the strength of the gravitational force between two bodies increases with their mass and decreases with the square of their distance apart.

Using this law of gravitation, Newton was able to show that the orbit of a smaller body, such as a planet, around a large body, such as the Sun, is an ellipse. Not only that, but knowing the speed and direction a planet was moving in, and how far it was from the Sun, he could calculate the precise shape of its orbit and predict exactly where it would be, as well as what speed and direction it would be moving in, at any time of his choosing. As you may guess, all of this comes in very handy when we are launching spacecraft on a jolly to the Moon.

IT REALLY ISN'T ROCKET SCIENCE

Now this may come as a disappointment, but the way a rocket works is actually pretty straightforward. There's the payload, which is the stuff you want to get up into space, like astronauts and buggies and national flags, and then there's the engine that gets them there. A rocket engine is basically made up of a

10 G is actually a very small number, 6.67×10^{-11} m^{3}kg^{-1}s^2 (roughly a tenth of a billionth of a m^{3}kg^{-1}s^2) because gravity is an extremely weak force. To make a reasonably large gravitational force like the one at the surface of the Earth, you need an awful lot of matter.

chamber with a nozzle. The fuel ignites in the chamber and fires hot gas out of the nozzle. As we know from Newton's third law, if the engine applies a force to a bunch of hot gas, in return that hot gas applies a force to the engine. And as we know from Newton's second law, when the force from the hot gas is greater than the weight of the rocket, it starts to accelerate and blasts off the launch pad. Once it's up in orbit, we know from Newton's law of gravitation that it will stay there without any further effort, tracing out a path in the shape of an ellipse. And from Newton's first law, any spaceship that escapes the Earth's gravitational field will simply chug off into space at a constant speed.

What else is there to know? Well, the fuel can be solid, like the ammonium perchlorate composite propellant (APCP) that powered the two enormous rocket boosters on the Space Shuttle, or liquid, such as kerosene mixed with liquid oxygen. Boosters that use liquid propellants tend to be safer, because they can be shut down if something goes badly awry, whereas once you've lit some solid propellant there's no turning back.[11]

Another thing to note is that, unlike with a jet engine, where the air intake supplies plenty of oxygen from the atmosphere to burn the fuel, rocket engines bring their oxygen with them. In the case of a solid fuel, the oxygen is provided by a solid chemical, such as ammonium perchlorate in the case of APCP or nitrate in the case of gunpowder. For a liquid fuel, the fuel is held in one tank while liquid oxygen is held in

---

11 A 'rocket' firework is basically a tiny rocket engine with solid propellant in the form of gunpowder, and as we all know once you light the blue touchpaper it either works or it doesn't. If there are a large group of people watching, it generally doesn't.

another, and a pumping system dumps them both in the combustion chamber where they can be ignited. Either way, the rocket is able to keep burning fuel even when it has left the atmosphere.[12]

Then there's a bit to know about rocket orbits. The first mistake most people make is in thinking that you need to burn fuel all the way from the Earth to the Moon, but of course you don't. You sort of hitchhike there on the Earth's gravitational field, burning most of your fuel to get up into Low Earth Orbit (LEO). Once there, you check your systems over and wait for your window of opportunity to transfer to the Moon. A short burn can then put you into a large elliptical orbit with the Earth at one end and the Moon at the other; in the Apollo missions this process was known as Trans-Lunar Injection (TLI). When you reach the Moon, you make another short burn to slow you down into a circular parking orbit so that you can send a small craft down to the lunar surface and back.

Then when everyone's safely back on board, another burn will put you back into another large elliptical orbit that will take you to the edge of the Earth's atmosphere; get the approach angle shallow enough and you won't bounce off the atmosphere back into space again. Then you splash down in the sea.[13]

And that's it. You know pretty much everything there is to

12 In the case of the Apollo missions, where the final two stages of the Saturn V rocket burned liquid hydrogen and oxygen, the combustion process even provided water for the astronauts to drink. Another reason I'll never make it to the Moon: I can't stand fizzy water.

13 As well as the major burns that change orbit, I should mention that there are also smaller burns for housekeeping tasks like separating the stages of the rocket, settling the fuel in the tanks before lighting up a new stage and making sure the whole thing stays on target while transiting between Heavenly Bodies.

know about the science of the Apollo missions. There are some subtleties I've left out, of course; neither the Earth nor the Moon is a perfect sphere, for example, and you also have the gravitational force of the Sun to contend with so you need to make a few tweaks to your calculations to get everything shipshape.[14]

Nevertheless, Newton's three laws of motion and his law of gravitation are the mathematical tools by which we calculated pretty much everything that mattered: what force it would require to achieve lift-off, how much fuel would be needed, where to launch the rocket, exactly where it would land on the Moon and where the capsule would land after re-entry into the Earth's atmosphere.

Right. You've earned your spurs; now's your chance to get to grips with the greatest rocket mankind has ever built: the Saturn V.

ROCKET MAN

All things being equal, a rocket probably wouldn't be your first choice for regular trips to the Moon. A more sustainable plan, for example, might be something along the lines of a large space station parked in Low Earth Orbit, which in turn acts as a launch base for other shuttle craft. Back in the 1960s when the race to the Moon was being run in earnest, of course, we

---

14 By the by, one of the great challenges of landing on an asteroid is that it doesn't have a regular gravitational field, which is one of the things that makes that particular NASA mission so exciting.

didn't have any large space stations, or any of the technology needed to build them. What we did have was rockets.

Or, to be more accurate, we had missiles. Dress it up how you like, the Soviet and American space programmes basically involved taking ballistic missiles based on the impressive V2 rocket that had been developed by the Nazis in the Second World War, and, rather than sticking a bomb in the nose, putting some astronauts in there instead. The fall of Germany in the spring of 1945 had seen both the Soviets and Americans in a desperate race to get to the V2 first; the Soviets got the rockets but the Americans ended up with the V2's designer, Werner von Braun.

Luckily the Soviets had their own master designer in the shape of one Sergei Korolev, and with the knowledge gained from the recovered V2 he successfully built the first staged rocket, the R-7, with which he launched the satellite Sputnik on 4 October 1957 and effectively fired the starting gun on the Space Race. The R-7 had two stages, both burning kerosene and liquid oxygen: staging is a vital innovation in rocket design, partly because it means you can shed unnecessary mass and partly because the fuel pumping system in later stages can be purposely designed to operate in the vacuum of space rather than under atmospheric pressure.

The Americans responded to Sputnik by setting up NASA – the National Aeronautics and Space Administration – and putting von Braun in charge of designing a rocket that could compete with the R-7. The family of rockets he invented are known collectively as Saturn; the one that took us to the Moon was the Saturn V. And what rockets they were.

I'm not usually one for facts and figures, but the Saturn V

really is worth rubbing your trousers over. It weighed 3,000 metric tonnes on the launch pad, most of which was fuel. The first of its three stages was made up of five enormous F-1 engines burning kerosene and liquid oxygen, and by the time it was discarded two minutes later the rocket was about 40 miles up in the air. The second stage, equipped with five J-2 engines, burned liquid hydrogen and oxygen for about six minutes and took the rocket to around 120 miles above the Earth. The third stage, made up of a single J-2 engine burning hydrogen and oxygen, burned for about two minutes to get the Saturn V into a circular Low Earth Orbit, then shut down while everyone got their breath back, then lit up again for about five minutes to put the whole thing on course for the Moon.

And what of the Russians? Sadly, Korolev's answer to the Saturn V, the N-1, never made it past the testing stage. Korolev's death in 1966 was a contributing factor, as was the complexity of the rocket he designed. Standing just a few metres shy of the Saturn V, the N-1 was also made up of three stages, the first stage of which had no less than thirty engines. Four test launches ended in failure, and as the Americans made landing after landing from July 1969 onwards, the Russians effectively conceded the Space Race.

THE DARK AGES IN EUROPE, 1972–2008

OK, so let's meet this one head-on. The Americans went to the Moon. The glory that was the lunar landings was part of a concerted effort that had begun with the Mercury programme where single astronauts got the hang of Low Earth Orbit, and which was then consolidated with the Gemini programme, which concentrated on docking in outer space, and was then focused into an all-important moonshot by the Apollo programme, which first swung past the Moon using its gravity as a slingshot with Apollo 8, and then finally landed the first craft on 20 July 1969 with Apollo 11. Throughout the six Moon landings, the Americans beamed TV pictures back to Earth, took photos, collected rocks and even set up reflectors, which we still fire lasers at today in order to measure their distance from the Earth.

As to the basic technology, as we've seen, that had been around since the Second World War with the invention of von Braun's V2 rocket. There were some difficult engineering challenges, of course; scaling up to the Saturn V was no simple task, nor was the design of the command and lunar modules and the navigation system. But it's not as if there is anything mysterious about the way that the US got there, or any complicated reason that the Russians failed; the simple truth of it is that the Americans managed to build a really, really big working rocket and the Russians didn't.

Yet the myth persists that mankind didn't go to the Moon. Why? One reason has to be that after Apollo 17 we haven't been back. After a while the whole thing began to feel a little less tangible, a little less real. Another is that the Watergate scandal, which came hot on the heels of the Moon landings and which involved the very president who had shaken the astronauts' hands at so many ticker-tape parades, did little to confirm the public's faith in American government and its agencies such as NASA. Yet another is CGI. For the live television audience of 1969, seeing was believing, whereas today audiences are used to the idea that images are just sets of numbers that can be manipulated at will.

And then there's the very nature of science itself. Science, by and large, works by a system of peer review. Before your paper is published in a recognised journal, it is scrutinised – and either accepted, corrected, or rejected – by a panel of experts. Once published, it stands until someone supplies further peer-reviewed evidence that contradicts it. You don't need to argue your case, or start a PR campaign, or cast aspersions on the motives of those who disagree with you. Unlike the humanities, science simply doesn't have a culture of argument and rebuttal. In 1931, for example, a pamphlet that was critical of relativity was published in Germany with the title 'One Hundred Authors Against Einstein'. Einstein famously replied, 'To defeat relativity one did not need the word of one hundred scientists, just one fact.'

For the majority of professional scientists, the idea that the Apollo missions were faked is so plainly at odds with the facts that they simply can't take it seriously. To even engage with such an evidentially challenged argument seems pointless, like having a conversation in French about the non-existence of France. Luckily, however, I am here as your middleman, and I am more than willing to stoop low enough to take on the naysayers. So let me briefly break ranks and take a few moments to address some of the conspiracy theorists' common arguments before hunkering down to the really fascinating stuff of just what there might be Out There for us to discover, and what the next leg of the journey is going to be.

If Neil Armstrong was the first man on the Moon, who was holding the camera?

No one. There was a camera mounted outside the lunar landing module. And I put it to you: if NASA are cunning enough to fake the entire Moon landings, would they really slip up by having a camera crew film Neil Armstrong stepping down from the lunar module? And wouldn't Stanley Kubrick, if he really was the director, have thought of that?

The astronauts would never have survived the Van Allen radiation belts.

Now there's some interesting science here. The Van Allen radiation belt is a doughnut-shaped cloud of charged particles out in space wrapped like a ring around the Earth's magnetic axis. The belt was first discovered in 1958 by Explorer 1, the first American satellite, and is created when high-energy protons and electrons from the solar wind and from cosmic rays get caught up in the Earth's magnetic field. It was detected by equipment designed by the American physicist James Van Allen and so bears his name. It's probably more familiar to you by one of its freak effects; sometimes there are disturbances in the Earth's magnetic field, where charged particles are whipped up out of the belt and flung towards the poles. When they hit the upper atmosphere, they strike gas molecules that then fluoresce, causing the Northern and Southern Lights.

NASA knew about the Van Allen radiation belts; they had discovered them, after all. As a result, the TLI burn that injected the Apollo craft into a moonward orbit was angled to clip the top of the Van Allen radiation belt rather than plough straight through the middle, while the return trip clipped the underside of the belt. Several of the Apollo astronauts experienced flashes within their eyes as they passed through the belts. The crew were constantly monitored for the radiation dose they received, and the total amount they ended up with was a hundred times less than that for radiation sickness.

You can't see stars in the photos of the astronauts.

That is because the landings were scheduled to take place in lunar daytime, when the Sun was shining. The photos are exposed for lunar sunlight, not for the much fainter starfield. Again, if a Space Agency mocked up the Moon landings, it would have been a bit of a blunder to forget to put in stars.

You can see the flag fluttering in some of the video footage but there's no wind on the Moon.

That isn't suspicious at all, now that we know from Newton's first law that stuff likes to keep quietly doing its thing at a constant speed unless hassled by a force. With no atmosphere to dampen the movement of the flag, it kept moving after it was jammed in place.

NASA made the Moon rocks.

Now you're just being silly . . .

CLOSE ENCOUNTERS

So 20 July 1969 was the first time that we ventured out from our home planet and set foot on our nearest neighbour, the Moon. The coming decades, as we shall shortly see, will bring other adventures as we first revisit the Moon, then possibly an asteroid, then no doubt travel to Mars and beyond. So what awaits us? Is there life on other planets in the solar system? Are there other civilisations out in the galaxy? And what about further out in the observable Universe, that dark sphere some 93 billion light years across?[15]

Let's be honest: when we wonder what we might find when we venture out of Earth's orbit to explore the solar system and beyond, we are not pondering what kind of rocks we might bring back. We are picturing something a bit more like *Star Trek*, where we cruise sedately from one star system to another, dispensing wisdom to suspiciously human-looking aliens dressed in skimpy space costumes. But what evidence do we have that aliens exist? And what would they look like? The Klingons? The Borg? It's time to ask one of the most intriguing questions in science, the one famously posed by the Italian physicist Enrico Fermi: where is everybody?

---

15 Because the Universe is expanding, although it is only 13.7 billion years old (only!), the edge lies some 47 billion light years away. At the time of writing, the oldest object we have seen is the galaxy known as UDFy-38135539, found by the Hubble Space Telescope, which was presumably named after someone's favourite bar code. The light from it has taken 13.1 billion years to reach us, but the galaxy itself is now some 30 billion light years away.

A HOUSE MADE OF BREADCRUMBS

As I drove my son to school this morning, one of my favourite DJs, the great Christian O'Connell, was holding forth in his morning slot on the British radio station Absolute FM. After trying to get a listener to come and fix a faulty Velux window in his home, then pondering as to why women keep nail varnish in the fridge, he asked a question that suddenly caught the physicist in me by the short and curlies. How many breadcrumbs would you need to build a house?

In scientific circles, this type of question is known as a Fermi question. Enrico Fermi, of course, is most famous for his work in radioactivity. You may already know that in 1942 he created the very first nuclear fission reactor in a basement under the football field at the University of Chicago, for example, or that he was instrumental in the Manhattan Project that beat the Nazis in the race to create the first nuclear bomb. But you may not be aware that, in the world of physics, Fermi is famous for another reason: his great skill in estimation.

Fermi loved to pose a certain kind of light-hearted scientific question. Famous examples include such teasers as 'How many atoms of Caesar's last breath do you inhale with each lungful of air?' and 'How many piano tuners are there in Chicago?'[16] Answering them is not only entertaining in its own right, but it fosters a real feel for the physical world. For

16 If you're interested, you breathe about one atom of Caesar's last breath every time you inhale, and there are about one hundred piano tuners in Chicago.

example, eyewitnesses even say that Fermi, who was at the Trinity test of the first atomic bomb in the Jornada del Muerto desert, New Mexico, on 15 July 1945, was able to accurately calculate the energy of that first explosion by waiting until after the flash, then dropping some small pieces of paper and seeing how far they were carried by the shock wave.

The way you answer a Fermi question is by making educated guesses based on things you already know about the world. If asked, 'What is the length of the equator?', for example, you might remind yourself that to fly across the United States you pass through four time zones, and travel a distance of roughly 3,000 miles. So each time zone must be about 750 miles, and since there are twenty-four time zones around the Earth, the equator must be roughly $24 \times 750 = 18,000$ miles long. I know, I know: the United States doesn't sit exactly along the equator and isn't exactly 3,000 miles across, but that's sort of the point. The answer will be right to within a factor of ten, or, as a physicist would say, within an 'order of magnitude'.[17]

The year 1950 saw a spate of flying-saucer sightings. Such events had been on the increase since the late 1940s, all conforming to roughly the same description: shiny disc-like spaceships, roughly the same size as man-made aircraft, that flew silently at extremely high speeds leaving no vapour trail. That summer, Fermi visited the Los Alamos National Laboratory in New Mexico, where the nuclear physicist Edward Teller was

17 In fact, I've just checked and the equator is 25,000 miles long, which isn't bad.

researching the feasibility of a hydrogen bomb.[18] As he, Teller and two others went to lunch, talk turned to the recent reports of flying saucers, Fermi's belief that they weren't really alien craft and the possibility of faster-than-light travel.

After all, reasoned Fermi, if aliens are ever going to play 'knock and run' with the hapless inhabitants of Earth, they are going to need a way of getting here, and with the vast distances between stars their spaceships would need to be extremely nippy. Teller's view was that there was a one in a million chance that, before the end of that decade, there would be evidence for an object travelling faster than the speed of light. Fermi was less pessimistic, putting the odds at around one in ten. According to Teller, the conversation then drifted to other more mundane subjects, like how to destroy humanity with nuclear weapons.[19] Then in the middle of lunch and seemingly apropos of nothing, Fermi suddenly asked, 'Where is everybody?'

His colleagues immediately grasped that he wasn't talking about the emptiness of the Fuller Lodge canteen. Fermi had asked himself the Fermi question, 'What is the possibility that alien civilisations exist?' and had been extremely puzzled by the answer. Having a rough estimate of the number of alien

18 The hydrogen bomb is a step on from the simple atomic bomb, which just uses runaway fission of uranium to create an explosive device. It was atomic bombs that were dropped on Hiroshima and Nagasaki at the end of the Second World War. The hydrogen bomb, on the other hand, uses hydrogen fusion together with uranium fission to create a truly monstrous explosion. It remains the blueprint for all nuclear warheads; when we say that this or that nation has 'got the bomb' it is the hydrogen bomb, or 'H-bomb', or sometimes 'thermonuclear bomb', that we are talking about.

19 All right, he didn't quite say that.

civilisations in the galaxy, he realised we ought to have been visited many, many times over. So where were they?[20]

WHAT ARE THE CHANCES?

Fermi's question to Edward Teller at Los Alamos has become a milestone in our quest to find other alien civilisations, so much so that it has gained notoriety as 'Fermi's Paradox'. Stated simply, the paradox is this: assuming that there's nothing that special about the Earth or the human species, intelligent life should be commonplace throughout the Universe. But where is the evidence?

To most people, it seems absurd to think that the human species is alone. One glance on a summer's night at the clot of stars in the heavens – some 100 billion just in our galaxy, and some 100 billion galaxies in the observable Universe – and it seems extraordinarily arrogant to assume we are in any way unique. In fact, it seems perfectly possible that somewhere out there are creatures identical to us, struggling to pay bills and find decent parking spots, let alone all sorts of weird

---

20 Fine, you might say, but how many breadcrumbs would you need to build a house? Well, I'm going to make my house a cube with a flat roof, with two large internal walls and two large internal floors, one at ground level and one at the first-storey level, so that the whole thing is a four-up, four-down. Let's say the house is 6 metres high, which looks about right to me, judging from the one I can see out of my study window. Counting the internal floors, the roof, the two internal walls and the four external walls, and assuming they are all 20 cm (0.2m) thick, means the total volume of building material I need is $9 \times 6 \times 6 \times 0.2 = 64.8$ m^3. Let's say that the average breadcrumb is a cube of side 2 mm, so that its volume is 8mm^3, or 8×10^{-9}m^3; the total number of breadcrumbs needed is therefore $64.8/(8 \times 10^{-9})=$ or 8.1 billion.

aliens with hyperdrives and effortlessly unbureaucratic Inter-Galactic Councils.

But where are they? The usual round of flying saucers, crop circles and little grey men have too human a fingerprint for them to be taken seriously as proof of intelligent alien life. We can't rule out the possibility that they are genuine, of course, but until we receive a single solitary piece of evidence – a wing mirror made of some extraterrestrial alloy, maybe, or some close contact that doesn't involve penetrative sex – we are going to have to face the fact that, so far as we can tell, no one has come to call. And, to get to the point, Fermi's Paradox is about more than a few flying saucers; as we shall see, if you run the numbers, we should be pushing past aliens on the high street. So where is everybody?

THE SEARCH FOR EXTRATERRESTRIAL INTELLIGENCE

In 1960, ten years after Fermi's famous conversation with Edward Teller at Los Alamos, an American physicist named Frank Drake decided it was time to find out. His plan was to target individual stars with a radio telescope to look for signals from extraterrestrial civilisations. To get the ball rolling, he asked himself the following Fermi question: how many alien signals should I expect to pick up with my radio telescope?

He expressed his answer as an equation and presented it at the first SETI conference at the National Radio Astronomy Observatory in Green Bank, West Virginia. It has since become known as the Drake Equation, but it's really just a back-of-an-envelope answer to a Fermi question. Written down, it looks a little bit formidable, but don't be downhearted because, as we shall see, it's really all rather straightforward and, more to the point, it is positively mind-expanding in the way that we think of life, the Universe and the life forms it may or may not contain. So here it is:

$$N = N^\star \times f_p \times n_e \times f_l \times f_i \times f_c \times f_L$$

I know it looks like something from Albert Einstein's blackboard, but it's really just a bunch of numbers multiplied together or, to be precise, two numbers ($N^\star$ and n_e) and five fractions (f_p, f_l, f_i, f_c and f_L). All it's saying is that to work out the number of alien civilisations − N − that might be transmitting a radio signal, you just need to work out how many of the stars in the galaxy are capable of supporting intelligent life.

COUNTING ALIENS

So to work out N, the number of alien civilisations that are sending out radio signals, we first need to work out how many stars are there in the galaxy, $N^\star$. Well, for my answer to Drake's Equation, I'm going to go for a conservative 100 billion. Current estimates run anywhere from 100 billion to 400 billion, so I'm well within my rights.

Next we need to ask ourselves, what fraction − f_p − of those

100 billion stars have planets? When I was at university in the eighties, there was no proof that other planets existed outside our own solar system. That all changed with the discovery of the very first planet in 1992, or rather the first two planets, since they are both orbiting a neutron star with the catchy name PSR 1257+12. Since then hundreds of others have been discovered, and our best guess is that 50 per cent of the stars we see in the galaxy have planets. So far, then, we have $N\star$=100 billion and f_p=0.5.

So what's next? Well, we've worked out what fraction of the stars in the galaxy have planets, so it's time to ask: of those stars with planets, how many of those planets are habitable? Habitable, by the way, generally means that the planet is the right distance from its home star to have liquid water on its surface. Life might equally start on a planet's moon, so our guess should take that into account too.

We do have some data on habitable planets, but not much. In fact, the first habitable extra-solar planet was discovered only in December 2011. It is orbiting a star called Kepler-22, after the Kepler Space Telescope that found it, about 600 light years away from Earth.[21] As we don't have much to go on, we can't do

---

21 NASA's Kepler Space Telescope finds planets by monitoring the brightness of stars. When a planet passes in front of its home star, it causes a tiny dip in the amount of light reaching the telescope. Remember Cygnus, the constellation that contains Deneb, one of the furthest stars visible with the naked eye? That's the constellation it's pointed at, partly because that region lies on the Milky Way and so there are lots of stars to look at, and partly because being pointed in that direction means that the Sun never shines into it. For extra points, you may like to know that the telescope orbits the Sun rather than the Earth, so that nothing in our solar system (planets, asteroids, Kuiper Belt objects, etc.) gets in the way. If you Google 'Habitable Exoplanets Catalog', you will be able to keep tabs on how many Earth-like planets it has turned up so far. At the time of writing there were four with a further twenty-three possible candidates.

much worse than look at our own solar system, where the Earth, the Moon and Mars are arguably all habitable to some degree, whereas you wouldn't want a holiday home on Venus or a minibreak on Jupiter. One habitable planet per solar system feels to me to be too pessimistic, and three feels like a step too far the other way, so I'm going to go for $n_e = 2$: on average, for every star with planets, there are two planets that can support life.

So far so good. Now we get to the more tricky guesses. Of those stars that have habitable planets, how many of those planets actually have life? A vital clue here is how early life on Earth took hold. If it got started quickly, we might assume life is a relatively common factor. If it took a while, we might guess that it's more of a rarity. Biologists generally agree that the first undeniable evidence of life is to be found in the rocks of the Pilbara Hills of Western Australia, where pillows of cyanobacteria flourished in shallow seas some 3.5 billion years ago.

But cyanobacteria are a relatively advanced form of life. So when did the very first organisms appear and where? Again, there are no straight answers, and everyone has their pet theory, from thermal vents in the seabed to warm puddles of fresh water. There is a tantalising hint of life in some 3.85-billion-year-old rocks from Akilia Island in Iceland, so I'm going to go for that as a start date. Seeing as life got started so quickly on Earth, I'm going to say that the chances of there being life on a habitable planet are roughly 100 per cent. So the number I'm going to plug in for f_l, the fraction of habitable planets where life takes hold, is 1. Again, there's no copper-bottomed probability theory or evidence behind this, but it feels to me like the right 'order of magnitude'.

Right. Nearly there. So of those planets where life starts, how many develop intelligent life? At this point we are swinging in the breeze with hardly a fact to our names. There are those who say intelligence is rare; after all, of all the species on Earth we are arguably the only one that has developed it. Then there are those who say evolution is bound to produce increasing complexity, and intelligence is simply a by-product of a complex organism. I've got to put something down, so I'm going to say, not that likely, but also not impossible: $f_i = 0.001$, or, in other words, the chance of intelligence on a planet that has life is about 0.1 per cent.

Let's move on the next term in Drake's Equation. Of those planets that develop intelligent life, what fraction $-f_c$ – will produce a civilisation capable of communicating with radio waves? Your guess is as good as mine, but it seems to me rather likely that if you develop intelligence you will explore the laws of the Universe, and we humans got to a theory of electromagnetism fairly sharpish once we began science in earnest. So perhaps again I should go for not-that-likely-but-not-impossible-yet-not-so-impossible-as-intelligence: something like $f_c = 0.1$.

Finally, we need to hazard a guess at f_L, the fraction of the galaxy's lifetime for which the average civilisation sends out radio signals. Now I'm an optimist in this regard. There are always those who claim the human race is facing imminent destruction, but it seems to me that once you reach the stage we happen to be at, you have enough technology and creativity to outlast pretty much any calamity. There's always the odd meteorite, of course, such as the one that sent all the dinosaurs home from the party without so much as a piece of cake, but they seem to come along every 100 million years or so and I don't think we'll last out anything like that long. So taking the age of the galaxy to be around 13 billion years, and the length of the average civilisation to be 10,000 years, the fraction of time that the average alien culture signals for will be 10,000/13 billion.

So I can now calculate N, the number of stars where I should be able to detect a radio signal:

$$N = N^\star \times f_p \times n_c \times f_l \times f_c \times f_p \times f_L$$
$$N = (100 \times 10^9) \times 0.5 \times 2 \times 1 \times 0.001 \times 0.1 \times (10,000/13 \times 10^9)$$
$$N = 7$$

In other words, there are about seven alien civilisations in our galaxy waiting for a call.

ET PHONE NO ONE

So what have we learned? Well, first, that it's extremely hard, given our present level of knowledge, to get a firm figure for the number of communicating alien civilisations in the galaxy. Nevertheless, given a reasonably optimistic set of assumptions, we would expect something of the order of ten signals, and even more if you believe that intelligence is common and that civilisations are long-lived. This, or something like it, would have been the calculation that Enrico Fermi made at Los Alamos back in 1950, and it is the crux of the Fermi Paradox.

Because after fifty years of listening we have yet to find anyone home.

Well, almost. There was the infamous 'Wow!' signal back in 1977, when one Jerry R. Ehman, working on the Big Ear radio telescope at Ohio University, detected a 72-second blast of radio waves coming from a point within the constellation of Sagittarius. Sadly the signal was not repeated, and many searches with much more powerful telescopes achieved a null result. What's more, SETI's searches have become more and more sophisticated, and now piggyback on a great number of the world's radio telescopes, continuously searching the heavens. In fact, thanks to the SETI@home program, your own humble laptop can do a bit of number-crunching in the search for alien intelligence; mine is analysing data for SETI even as I type, and I strongly urge you to hook yours up to do the same.

The problem is, of course, there's a lot of galaxy out there. Even if you believe there are 10,000 broadcasting civilisations in the Milky Way, you've still got your work cut out finding them among 100 billion stars.[22] To make matters worse, there's the whole question of what frequency the aliens are broadcasting at, not to mention the question of whether they are broadcasting radio waves at all, rather than, say, firing laser beams or blasting out gamma rays. Maybe the signals are everywhere, but we're just looking in the wrong places.

After all, who is to say that an alien civilisation would send out signals in the first place, or whether they would fit our definition of intelligence, or even be carbon-based life forms

22 One signal per 10 million stars by my reckoning.

living on an Earth-like planet? What if they aren't organisms at all in the biological sense? Might they not be so advanced that they no longer inhabit the physical Universe, preferring instead to upload themselves into a virtual plane of their own choosing where they spend all day playing Mario Kart?

SO, COME ON, WHERE IS EVERYBODY?

For me, this has been the perfect topic to end this book. To even begin to answer the question of whether there is intelligent life elsewhere in the galaxy, we need to draw on so many glorious past discoveries as well as relish the prospect of so many glorious discoveries yet to come. By exploring the origins of life we can begin to build a picture of how it might take hold on other planets in other solar systems; by exploring the origins of matter we have our best chance of keeping the lights on and one day paying them a visit. Although, as we have learned, for our species it has never been about the destination. From the first clutch of souls to leave Africa and settle the Old and New Worlds, it has always been about the journey: the thrill of exploration for its own sake.

And my own answer to the Fermi Paradox? Well, I reserve the right to change my mind, but this is where I am at the moment: we are alone. At first, hearing that might sound pessimistic, but it really isn't. My logic is this: the same physical laws are at work, whether you live on Kepler 22b or on Earth. Somewhere out there on some neglected extra-solar planet, snow is softly falling and waves are lapping tropical shores. The striking feature of, say, Mars or the Moon is not how alien they are, but how oddly familiar, how like a forgotten corner of some American National Park. We may yet find microbes in some deep lunar trench, or under a shady rock on Mars, and it seems perfectly likely that there are gazillions of similar rocky planets where low-level life would develop, given a puddle of water, 4 or 5 billion years and the odd lightning strike.

Biology, it seems to me, may be commonplace, but intelligence? Frankly, I'm not even sure I've got it. Hard as it is to stomach, evolution just doesn't care a hoot about whether you can do the *Times* crossword; it just cares about whether or not you pass your genes on to your offspring. On those terms, the nematode worm has us beaten hands down; they outnumber us something like a billion to one. There is no great advantage, really, to our intelligence in survival terms; some might say it's a handicap, given our ability to invent ever more colourful ways of destroying one another.

No, I think we are alone in the galaxy; the aliens aren't here because there aren't any. But I don't find that depressing, in fact, quite the opposite; I find it extremely uplifting. Because, in my view, *Star Trek* didn't have it too wrong after all. After all, we humans have got form. Like many times before in our history, a small group will strike out from home, looking for adventure. They will spread, like Darwin's finches, to nearby planets, evolving into new human-like species as they go. And one day, perhaps, some spaceship from planet Earth may even journey among them, dispensing wisdom and ogling them in their skimpy bikinis. We are the Borg, the Klingons and the Romulans. We just haven't set off yet.

FURTHER READING

If I have one wish for this book, it is that it acts as a kind of 'gateway' drug to the intoxicating world of popular science. By way of introduction, here is some of the good stuff . . .

GENERAL READING

Bad Science by Ben Goldacre, Harper Perennial, 2009

Physics for Future Presidents: The Science Behind the Headlines by Richard Muller, W. W. Norton & Co., 2008

PARTICLE PHYSICS

Dreams of a Final Theory: The Search for the Fundamental Laws of Nature by Steven Weinberg, Vintage, 1993

Surely You're Joking, Mr Feynman: Adventures of a Curious Character as Told to Ralph Leighton by Ralph Leighton, Richard P. Feynman and Edward Hutchings, Vintage, 1992

ASTROPHYSICS AND RELATITIVY

Relativity: The Special and the General Theory by Albert Einstein, Pober Publishing Company, 2010

Why Does E=mc²? (And Why Should We Care?) by Brian Cox and Jeff Forshaw, De Capo, 2010

EVOLUTION

Your Inner Fish: The Amazing Discovery of Our 375-Million-Year-Old Ancestor by Neil Shubin, Penguin, 2009

Catching Fire: How Cooking Made Us Human by Richard Wrangham, Profile, 2010

DNA

Genome: The Autobiography of a Species in 23 Chapters by Matt Ridley, Fourth Estate, 2000

What Mad Pursuit: A Personal View of Scientific Discovery by Francis Crick, Basic Books, 1990

COOKING

Cooking for Geeks: Real Science, Great Hacks and Good Food by Jeff Potter, O'Reilly Media, 2010

Molecules at an Exhibition: Portraits of Intriguing Materials in Everyday Life by John Emsley, Oxford Paperbacks, 1999

CLIMATE

The Weather Book: Why It Happens and Where It Comes From by Diana Craig, Michael O'Mara, 2009

The Little Ice Age: How Climate Made History 1300–1850 by Brian M. Fagan, Basic Books, 2001

SPACE TRAVEL AND ALIENS

How Spacecraft Fly: Spaceflight Without Formulae by Graham Swinerd, Springer, 2008

The Eerie Silence: Searching for Ourselves in the Universe by Paul Davies, Penguin, 2011

ACKNOWLEDGEMENTS

First thanks must go to Elly James at hhb, who cold-called me about five years ago after hearing me bang on about physics on Radio 4 and cajoled, cosseted and corralled me into getting all this down on paper. And equal first thanks to my fabulous publisher, Antonia Hodgson at Little, Brown, who has had such an unswerving passion for this book and such a crystal-clear vision of its future. A big shout out also to Hannah Boursnell, Sally Wray and Celia Levett and all the amazing team at Little, Brown and to the ever-supportive Heather Holden-Brown and Rob Dinsdale at hhb.

I'd never have been able to write about science without my truly gifted teachers: Mr Bailey at Willaston County Primary School; Mr Clarke, Mr Gee, Mr Roberts and Mr Davies at Malbank Comprehensive School, Nantwich; Dr Davies and Professor Shakeshaft at St Catharine's College, Cambridge; and Professor Pepper at the Cavendish. The mistakes are all mine and the inspiration is all theirs. Thanks also to Tom Haine, now Professor Haine of Johns Hopkins University, Baltimore, who was the constant object of my admiration and a ready source of solutions to Green's Functions when we were undergraduates together.

I'd like to thank James Harding at *The Times* for encouraging me to write about science and I'd also like to express my gratitude to Giles Whittell, David Edwards and David Reay and all at *Eureka* magazine for publishing my column and in the process giving me my monthly sneak-peek at the finest science writing in any national newspaper.

I owe a huge debt to everyone who helped me with the manuscript. Jack Sandle was kind enough to read my first two chapters and offered some much-needed words of enthusiasm, Jobina Hardy has been an invaluable researcher and sounding-board for the biology chapters and Suzy McClintock did everything from generating graphs to locking me in my shed until I had finished my illustrations.

I am also indebted to the experts who were kind enough to cast a seasoned eye over my work. Dr Ian Strangeways corrected some of my basic misunderstandings of temperature measurement and Professor Mitchell at the Met Office saved me from some embarrassing blunders in meteorology. Likewise Dr Caren Brown rectified some of my dunderhead mathematical mistakes and the wonderful Dr Frances Astley-Jones put me right on some cock-ups in the biology chapters that would have made even a chiropractor blush.

I was also lucky enough to see many extraordinary scientific institutions at first hand. Dave Britton of the Met Office was kind enough to invite me to the Hadley Centre to see computer-modeling of climate; my heartfelt thanks to him and Sarah Holland for all their help and advice. CERN showed me round the Large Hadron Collider for my Radio 4 series *My Great Big Particle Adventure* and gave me access to some of the leading lights of particle physics; I am grateful to everyone who gave their time, but special mention must go to John Ellis for his lucid explanations of the Higgs field and String Theory. I also had the good fortune to visit the JET fusion laboratory in the process of filming my BBC2 *Horizon* programme *One Degree* and I would like to thank them profusely for answering my questions about the future of nuclear fusion and also for making sure I didn't touch any of the buttons.

The other 'thank yous' I want to make are more personal. Jessica Parker, my partner, raised our family pretty much single-handed during umpteen late-night deadlines and I am unfeasibly grateful and promise to make it up to her. One of the reasons I wrote this book is that I hope that one day our boys Sonny and Harrison might enjoy it. My mum, Marion, my sisters Bronwen and Leah and my nephews and nieces Billy, Jude, Beth and Dylan were a constant source of love and support. Bruce McKay, Ol Parker, Rob Bratby, Steven Cree, Jez Butterworth and Pierre Condou were staunch allies as always. Alexander Armstrong has endured half a lifetime of anecdotes about science for which I offer a formal apology. I also want to formally thank him for giving me a career in the first place.

Speaking of a career, I also wouldn't have one without Samira Higham and all at Independent, and Jimmy Mulville and all at Toff Media and Hat Trick Productions. I am indebted to Josephine Green, my redoubtable PA, and I would never have survived *Death In Paradise*, which I filmed in Guadeloupe at the same time as writing, without my assistant Kim Read. My great friend Imogen Edwards-Jones was, as ever, a font of sound advice in all matters literary and even found me somewhere to run away and write. Thank you too to David Mitchell and Brian Cox for being kind enough to read my finished chapters and to supply quotes when I think they both probably had quite a lot on their plates running the world.

ACKNOWLEDGEMENTS

Last-but-one, I want to thank all the inspirational scientists who have created enduring works of popular science. *Cosmos* by Carl Sagan, *The Ascent of Man* by Jacob Bronowski and Stephen Hawking's *A Brief History of Time* all fired my young imagination, and David Attenborough's and Brian Cox's peerless BBC series do us all a service, not just by making science accessible but also revealing its great beauty.

And last, but most certainly not least, I want to thank NASA for landing on the Moon.

PICTURE CREDITS